JUNIOR, INTERMEDIATE AND SENIOR MATH OLYMPIADS

JUNIOR, INTERMEDIATE AND SENIOR MATH OLYMPIADS

Roman Kvasov, Ph.D.

Published by 42 Points. San Juan, Puerto Rico, 2022.

CONTENTS

PART III   SENIOR MATH OLYMPIADS

Dedicated to all students who love Math Olympiads

INTRODUCTION

This book presents a collection of problems from a series of Junior, Intermediate, and Senior mathematics competitions, commonly known as the "42PMO" or "42 Points Math Olympiad". These competitions were organized by the author between 2017 and 2024. The problems come with detailed solutions and are structured as individual exams that can be approached in any order or utilized for practice. Some problems are well-known classics, but the majority are original, designed to cover topics in Math Olympiad algebra, geometry, number theory, and combinatorics.

It is noteworthy that in many countries, the junior math olympiads encompass grades 7-10, with grades 7-8 being considered part of the junior level. Drawing from the author's extensive experience working with students of varying ages and cultural backgrounds, a recurring observation is that younger students in grades 7-8 often face challenges competing with their older counterparts. This observation led to the recognition of the need for a separate category tailored to the capabilities and experiences of these younger participants.

In response to this, the author made the decision to introduce a category specifically designed for intermediate math olympiads, targeting students in grades 9-10. This adjustment allows for a more accurate representation of the skill level and mathematical maturity of participants in this age group.

Consequently, the junior math olympiads are now designated for students up to the 8th grade, offering a platform designed for participants who may have limited exposure to proof-based mathematics. This categorization not only enhances the inclusivity of the competition but also ensures that participants are challenged at an appropriate level, fostering a positive and rewarding experience for all involved.

Based on these observations, participants were classified into three levels of difficulty during the competition: Junior Level (7-8 grades), Intermediate Level (9-10 grades), and Senior Level (11-12 grades). This categorization aligns with the level distribution observed in some national Math Olympiads, such as the Australian Math Olympiad.

It is important to highlight that there are no proof-based math olympiads in the United States for the students of 7-8 grades. Therefore, the Junior Level competition is unique in a sense that it gives the younger students the opportunity to try the problems set in a proof-based format that, however, do not require much knowledge or experience of writing proofs.

The difficulty of most problems at the Intermediate Level is comparable to that of the Central America and the Caribbean Math Olympiad (OMCC), the Junior Balkan Math Olympiad (JBMO), and the Puerto Rico Team Selection Test (OMPR TST). This suggests that participants at the Intermediate Level can use the competition as a training ground for other contests. The more challenging problems at the Intermediate Level are similar in difficulty to those in the United States of America Junior Math Olympiad (USAJMO) and the Canadian Junior Mathematical Olympiad (CJMO).

The difficulty of the problems at the Senior Level is comparable to problems of the British Mathematical Olympiad (BMO) and the United States of America Math Olympiad (USAMO). Currently, the Senior Level is also the most challenging mathematics competition held in Puerto Rico.

Until 2019, the competitions took place at TASIS School in Dorado, Puerto Rico. Due to the COVID-19 pandemic, the 2020 competitions were held online, leading to increased participation from students in various countries, including the mainland United States, Canada, Venezuela, Switzerland, Spain and India. Since then, the competition has remained online and regularly attracts more than 60 participants each year.

All participants receive certificates of participation, while medals are presented to the upper 50% of participants. Among the most successful math olympians who have taken part in the competition are Jessica Wan, Enrique Rivera, Johann Williams, Nirvana Marrero, Rafael Gomez, Carlos Rodriguez, and many others.

The author would like to wish the reader the best of luck on their math olympiad journey and hopes that they will enjoy the book.

Roman Kvasov, Ph.D.

PART I

JUNIOR MATH OLYMPIADS

Junior Level math olympiads cater to students up to the 8th grade, providing an excellent opportunity for participants with limited exposure to proof-based mathematics. These competitions cover fundamental topics in algebra, geometry, combinatorics, and number theory, designed to challenge young minds and foster a deeper understanding of mathematical concepts.

In the realm of algebra, participants can expect problems involving operations on rational expressions, various types of factorization (common factor, difference of squares, sum and difference of cubes, grouping technique, Simon's Favorite Factoring Trick), techniques like completing squares and solving linear and polynomial equations in one variable. These foundational algebraic principles serve as building blocks for more advanced mathematical reasoning.

Geometry problems at the junior level often revolve around angle chase, congruent triangles, and properties of isosceles and equilateral triangles. The focus is on instilling a solid grasp of geometric principles and honing problem-solving skills.

Combinatorics introduces participants to basic counting methods, the concept of parity, the Pigeonhole principle, and simple coloring techniques. These problems not only test mathematical reasoning but also encourage creative thinking and logical deduction.

In the area of number theory, junior-level participants explore topics such as decimal representation, residues by a modulus, factorization in Diophantine equations, selecting the appropriate modulus for a given problem, employing casework strategies, and correctly applying the properties of divisibility and congruence. These problems delve into the fascinating world of number theory, engaging participants in the exploration of patterns and relationships within integer numbers.

To prepare effectively for Junior Math Olympiads, aspiring participants should focus on mastering these fundamental concepts. This can be achieved through a combination of regular practice, exposure to a variety of problems, and participation in mock exams. Developing a deep understanding of the basic principles in algebra, geometry, combinatorics, and number theory will empower students to approach olympiad problems with confidence and creativity. Additionally, seeking guidance from experienced mentors or teachers, and collaborating with peers in problem-solving sessions, can further enhance one's preparation for the challenges posed by junior math olympiads.

CHAPTER 1

JUNIOR LEVEL EXAM OF 2017

Problem 1 (Number Theory)

Find all ordered pairs of prime numbers (p, q), such that

$$p + q = 2019$$

Solution

Answer: $(2, 2017)$, $(2017, 2)$.

Notice that since 2019 is an odd number, then one of the numbers p or q is even, and, therefore is equal to 2. From here the other numbers is equal to 2017. It is not hard to check that 2017 is indeed a prime.

Problem 2 (Algebra)

Three nonzero real numbers a, b and c are such that

$$\frac{a}{b} = \frac{a + c}{b + c}$$

Prove that $a = b$.

Junior, Intermediate and Senior Math Olympiads
by Roman Kvasov, Ph.D.

13

Solution

Let us cross-multiply and rewrite the given equation as follows:

$$\frac{a}{b} = \frac{a+c}{b+c}$$
$$a(b+c) = b(a+c)$$
$$ab + ac = ab + bc$$
$$\cancel{ab} + ac = \cancel{ab} + bc$$
$$ac = bc$$

Since c is a nonzero real number, then we can divide both sides of the last equation by c and obtain $a = b$, as desired.

Problem 3 (Number Theory)

Let $S(n)$ be the sum of digits of the natural number n. What is the number n if $S(n) = 27$ and $S(n+1) = 1$?

Solution

Answer: $n = 999$.

It is not hard to see that since $S(n+1) = 1$, then $n+1$ is a power of 10. Therefore, n consists entirely of digits 9. Since $S(n) = 27$, then n has exactly three digits 9 and is equal to 999.

Problem 4 (Combinatorics)

Given seven positive integers. Show that two of them can be chosen whose sum of whose difference ends with 0.

Solution

Notice that if any two of these numbers end on the same digit, then their difference ends with 0. Let us assume that all numbers end on different digits and that no sum of any two numbers ends with 0.

Let us consider the following pairs of digits:

$$(1,9), (2,8), (3,7), (4,6), (5,5), (0,0)$$

Notice that the sum of the two digits in each pair ends with 0. It is clear that there cannot be more than one integer in each pair, otherwise the sum of these two integers will end with 0. Therefore the total number of numbers cannot be more than 6. We obtained a contradiction.

Problem 5 (Combinatorics)

Given a regular polygon with 2022 vertices, each containing the numbers 1 or -1, Jake takes each side and writes the product of the numbers written at its vertices. It turns out that the sum of the products on the sides is equal to zero. Show that he made an error in his calculations.

Solution

Let us assume that Jake did not make any errors in his calculations.

Notice that each of the numbers that Jake writes on the sides is either 1 or -1. Since the sum of all numbers on the sides is zero then there are exactly 1011 numbers 1 and 1011 numbers -1. This implies that the product of all numbers on the sides is negative. However, since each number written on the vertices participates exactly in two sides, then the product should be positive. We obtained contradiction.

Problem 6 (Geometry)

Equilateral triangles ABP, BCQ and CAR are constructed externally on the sides AB and BC of the triangle ABC. Prove that $AQ = BR = CP$.

Solution

The solution presented below refers to Figure 1.1.

Let us prove that $AQ = CP$.

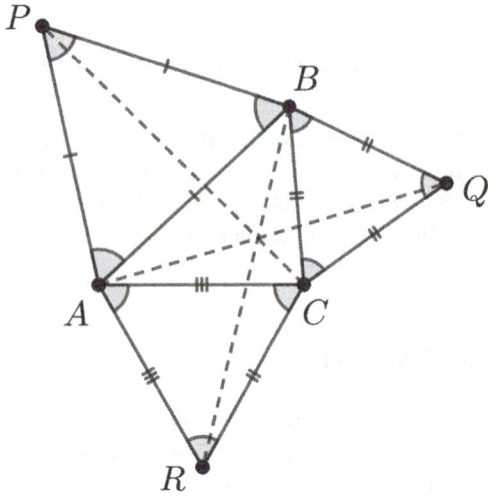

Figure 1.1   Triangles ABQ and PBC are congruent in Problem 6.

Start by noticing that $AB = BP$ and $BC = BQ$. Also

$$\angle ABQ = \angle ABC + 60° = \angle PBC$$

From here, the triangles ABQ and PBC are congruent by Side-Angle-Side Postulate. From here $AQ = CP$.

Similarly it can be shown that $AQ = BR$, from where $AQ = BR = CP$, as desired.

CHAPTER 2

JUNIOR LEVEL EXAM OF 2018

Problem 1 (Geometry)

In the triangle ABC it is given that $\angle ABC = 20°$ and $\angle BAC = 80°$. Prove that $AB = BC$.

Solution

The solution presented below refers to Figure 2.1.

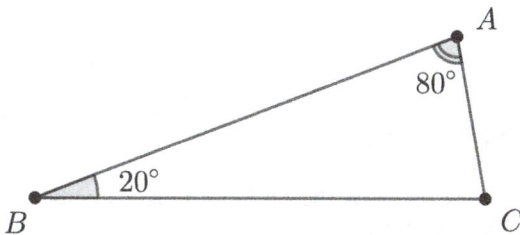

Figure 2.1   Triangle ABC is isosceles in Problem 1.

Notice that

$$\angle ACB = 180° - \angle ABC - \angle BAC = 180° - 20° - 80° = 80°$$

Junior, Intermediate and Senior Math Olympiads
by Roman Kvasov, Ph.D.

17

This implies that the triangle ABC is isosceles, and consequently, $AB = BC$, as desired.

Problem 2 (Combinatorics)

There is a pile of 2018 stones. Lucas and Summer are playing the following game: at each turn, a player can take either two or three stones from the pile. The loser is the player who cannot make a move. Which of the players has a winning strategy if Summer goes first?

Solution

Answer: Summer has a winning strategy.

On her first move Summer should take 3 stones. Then she should react to every Lucas' move as follows:

- If Lucas takes 2 stones, then Summer should take 3 stones.

- If Lucas takes 3 stones, then Summer should take 2 stones.

Notice that, after each Summer's move the number of the stones left in the pile is a multiple of 5 (these are the so-called "winning positions" of this game). Therefore, after Summer makes 404 moves the pile will have zero stones and Lucas will lose.

Problem 3 (Number Theory)

Let $S(n)$ be the sum of digits of the positive integer number n. Find the first two positive integers n, such that $S(n) = 2$.

Solution

Answer: $n = 2, 11$.

Let us assume that n is a positive integer, such that $n \leq 9$. Then we have that $S(n) = n$. This implies that $n = 2$ is the only number that works for $n \leq 9$.

Let us assume that n is a positive integer, such that $n \geq 10$. Then we have that $S(10) = 1$, while $S(11) = 2$.

Consequently, the first two natural numbers n, such that $S(n) = 2$ are $n = 2, 11$.

Problem 4 (Combinatorics)

Is it possible for a knight's move to traverse all cells of a chessboard, starting from bottom-left square and ending at the top-right square, visiting each cell exactly once?

Solution

Answer: no, it is not possible.

Notice that the bottom-left square is black (see Figure 2.2).

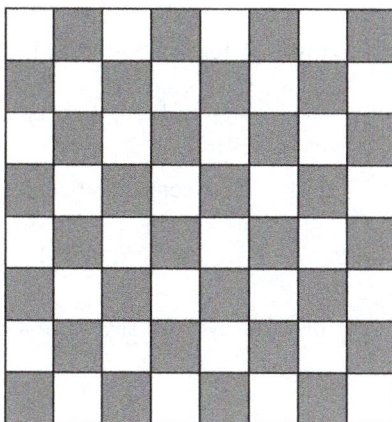

Figure 2.2   Bottom-left and top-right squares are black in Problem 4.

After each odd move, the knight is on a white square, and after each even move, it is again on a black square. To traverse all cells of the chessboard, 63 moves are required.

After each odd move, the knight is on a white square, and after each even move, it is on a black square. Therefore, on the 63rd move, the knight must land on a white square. However, the top-right square is black; therefore, the knight cannot end up in this square after the final move.

Problem 5 (Number Theory)

Show that $2^n + 1$ is not divisible by 7 for any natural n.

Solution

Start by noticing that
$$2^3 \equiv 8 \equiv 1 \pmod 7$$

Now we will work with the powers modulo 3 and proceed by doing the following casework:

- If $n \equiv 0 \pmod 3$, then $n = 3k$ for some positive integer k. Therefore

$$2^n + 1 \equiv 2^{3k} + 1 \equiv \left(2^3\right)^k + 1 \equiv (8)^k + 1 \equiv (1)^k + 1 \equiv 2 \pmod 7$$

- If $n \equiv 1 \pmod 3$, then $n = 3k + 1$ for some positive integer k. Therefore

$$2^n + 1 \equiv 2^{3k+1} + 1 \equiv 2\left(2^3\right)^k + 1 \equiv 2(8)^k + 1 \equiv 2(1)^k + 1 \equiv 3 \pmod 7$$

- If $n \equiv 2 \pmod 3$, then $n = 3k + 2$ for some positive integer k. Therefore

$$2^n + 1 \equiv 2^{3k+2} + 1 \equiv 4\left(2^3\right)^k + 1 \equiv 4(8)^k + 1 \equiv 4(1)^k + 1 \equiv 5 \pmod 7$$

We conclude that $2^n + 1$ is not divisible by 7 for any natural n.

Problem 6 (Algebra)

Real numbers x, y and z satisfy the equations:

$$xyz = 1$$
$$x + y + z = 1$$

Prove the equality

$$(xy + xz - x)(yz + xy - y)(xz + yz - z) = -1$$

Solution

Start by noticing that from the equation $x + y + z = 1$ we have

$$x + y - 1 = -z$$
$$y + z - 1 = -x$$
$$z + x - 1 = -y$$

Let M denote the expression on the left-hand side of the needed equality. Let us factor out the common factor in each parenthesis of the expression M. We have

$$
\begin{aligned}
M &= (xy + xz - x)(yz + xy - y)(xz + yz - z) \\
&= x(y + z - 1)\, y(z + x - 1)\, z(x + y - 1) \\
&= xyz(y + z - 1)(z + x - 1)(x + y - 1) \\
&= (1)(y + z - 1)(z + x - 1)(x + y - 1) \\
&= (-x)(-y)(-z) \\
&= -xyz \\
&= -1
\end{aligned}
$$

as desired.

CHAPTER 3

JUNIOR LEVEL EXAM OF 2019

Problem 1 (Algebra)

Three nonzero real numbers x, y and z are such that

$$x^2 - y = 0$$
$$y^2 - z = 0$$
$$z^2 - x = 0$$

Find all possible values of xyz.

Solution

Answer: $xyz = 1$.

Let us rewite the three given equations as

$$x^2 = y$$
$$y^2 = z$$
$$z^2 = x$$

and multiply them:

$$x^2 y^2 z^2 = xyz$$
$$(xyz)^2 = xyz$$

Junior, Intermediate and Senior Math Olympiads
by Roman Kvasov, Ph.D.

23

Since x, y and z are nonzero real numbers, then we can divide both sides of the last equation by xyz. Consequently, we obtain

$$xyz = 1$$

as desired.

Problem 2 (Number Theory)

Let $S(n)$ be the sum of digits of the number n. What is the smallest positive integers n, such that
$$S(n) + S(n + 10) = 11$$

Solution

Answer: $n = 5$.

Notice that $n = 1, 2, 3, 4$ do not work, since

$$S(1) + S(11) = 3$$
$$S(2) + S(12) = 5$$
$$S(3) + S(13) = 7$$
$$S(4) + S(14) = 9$$

Since the number $n = 5$ satisfies the conditions of the problem:

$$S(5) + S(15) = 11$$

then we conclude that the asnwer is $n = 5$.

Problem 3 (Algebra)

The numbers a, b and c satisfy the equations

$$a^2 + b^2 = c^2$$
$$b^2 + c^2 = a^2$$

Prove that $abc = 0$.

Solution

Let us start by subtracting the equations:

$$\left(a^2 + b^2\right) - \left(b^2 + c^2\right) = c^2 - a^2$$
$$a^2 + b^2 - b^2 - c^2 = c^2 - a^2$$
$$a^2 - c^2 = c^2 - a^2$$
$$2a^2 = 2c^2$$
$$a^2 = c^2$$

Now from the first equation we have

$$a^2 + b^2 = c^2$$
$$c^2 + b^2 = c^2$$
$$b^2 = 0$$
$$b = 0$$

and, therefore $abc = 0$, as desired.

Problem 4 (Combinatorics)

There is a pile of 19 stones on a table. You are allowed to perform the following operation: you choose one of the piles containing more than 1 stone, throw away one stone from that pile and divide it into two smaller (not necessarily equal) piles. Is it possible to reach a situation when there are 8 piles and 11 stones in total?

Solution

Answer: it is impossible.

Let us assume that it is possible to have exactly 8 piles with a total of 11 stones.

Then the sum of the number of piles and the number of number stones is

$$8 + 11 = 19$$

Notice that after the first operation we have 2 piles and 18 stones in total. After the second operation we have 3 piles and 17 stones in total. After the third operation we have 4 piles and 16 stones in total. Therefore, the sum of the total number of stones and the total number of piles is always 20 (this is a so-called "invariant" of this problem).

We conclude that it is impossible to obtain a situation where the total number of stones and the total number of piles is 19. We obtained a contradiction.

Problem 5 (Geometry)

Points D and G are chosen inside the triangle ABC, such that $\angle BDC = 150°$ and $\angle BGC = 100°$. It is known that $\angle ABG = \angle GBD$ and $\angle ACG = \angle GCD$. Prove that $\angle BAC = 50°$.

Solution

The solution presented below refers to Figure 3.1.

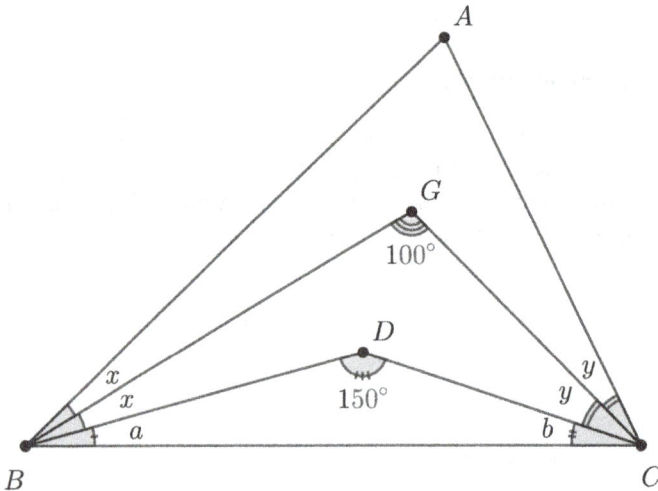

Figure 3.1   Points D and G located inside the triangle ABC in Problem 5.

First let us put $\angle ABG = \angle GBD = x$, $\angle ACG = \angle GCD = y$, $\angle DBC = a$, $\angle DCB = b$.

From the triangle BDC we have

$$a + b = 180° - 150° = 30°$$

From the triangle BGC we have

$$x + y = 180° - 100° - a - b = 50°$$

Therefore, from the triangle ABC we obtain that

$$\angle BAC = 180° - 2x - 2y - a - b = 180° - 100° - 30° = 50°$$

Problem 6 (Number Theory)

Two prime numbers are such that their product is greater than their sum by 43. Find all such numbers.

Solution

Answer: the only such numbers are 3 and 23.

Let the numbers be p and q. We have the following equation:

$$pq = p + q + 43$$

Let us factor it using Simon's Favorite Trick:

$$pq = p + q + 43$$
$$pq - p - q = 43$$
$$pq - p - q + 1 = 44$$
$$p(q - 1) - (q - 1) = 44$$
$$(p - 1)(q - 1) = 44$$

Notice that 44 can only be factored as $1 \cdot 44$, $2 \cdot 22$ or $4 \cdot 11$.

Without loss of generality, we will proceed by doing the following casework:

- If $p - 1 = 1$ and $q - 1 = 44$, then $p = 2$ and $q = 45$. However, 45 is not a prime. Contradiction.

- If $p - 1 = 2$ and $q - 1 = 22$, then $p = 3$ and $q = 23$.

- If $p - 1 = 4$ and $q - 1 = 11$, then $p = 5$ and $q = 12$. However, 12 is not a prime. Contradiction.

We conclude that the only such numbers are 3 and 23.

CHAPTER 4

JUNIOR LEVEL EXAM OF 2020

Problem 1 (Number Theory)

Let $S(n)$ be the sum of digits of the natural number n. Is it possible that $S(n) = 15$ and $S(n + 1) = 15$, for some $n \in \mathbb{N}$.

Solution

Answer: no.

Since $S(n) = 15$, then n is divisible by 3. However, this implies that $n + 1$ is not divisible by 3, and consequently, the sum of its digits cannot be equal 15.

We obtained a contradiction.

Problem 2 (Geometry)

In the triangle ABC, the point D is chosen on the side BC, such that $AD = BD$. It is known that

$$\angle DAC = \angle ABC - \angle ACB$$

Prove that $\angle ADC = 120°$.

Junior, Intermediate and Senior Math Olympiads
by Roman Kvasov, Ph.D.

29

Solution

The solution presented below refers to Figure 4.1.

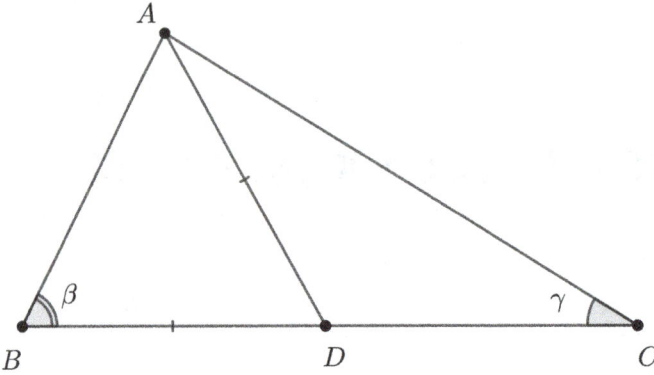

Figure 4.1   Triangle ADB is isosceles in Problem 2.

First let us put $\angle ABC = \beta$ and $\angle ACB = \gamma$.

Notice that we are given that $\angle DAC = \beta - \gamma$. The triangle ADB is isosceles and, therefore, we have that

$$\angle BAD = \angle ABD = \beta$$

Form here

$$\angle ADB = 180° - 2\beta$$

and

$$\angle ADC = 2\beta$$

From the triangle ADC we have

$$(2\beta) + (\beta - \gamma) + (\gamma) = 180°$$
$$3\beta = 180°$$
$$\beta = 60°$$

Since $\angle ADC = 2\beta$, then

$$\angle ADC = 2\,(60°) = 120°$$

as desired.

Problem 3 (Algebra)

Prove that there do not exist distinct positive real numbers a, b, x that satisfy the equations

$$(a + x)^2 = (b + x)^2 + 1$$
$$a^2 = b^2 + 1$$

Solution

Let us assume that such numbers exist. Let us subtract the equations:

$$(a + x)^2 - a^2 = (b + x)^2 + 1 - b^2 - 1$$
$$a^2 + 2ax + x^2 - a^2 = b^2 + 2bx + x^2 + 1 - b^2 - 1$$
$$2ax = 2bx$$
$$2x(a - b) = 0$$

Since $x > 0$, then from the last equation we have $a - b = 0$. From here $a = b$, which leads to a contradiction.

Problem 4 (Combinatorics)

6 students participated in a Math Club. During the meeting each student talked with every other student. It is known that they all talked about two different topics and each pair of students only talked about one topic. Prove that there are three students that talked about the same topic among themselves.

Solution

Let us assume that no three students talked about the same topic among themselves. Let T_1 and T_2 be the two topics that all students talked about. Let us pick a random student A. Since A talked to other 5 students, then by the Pigeonhole Principle there exists 3 students with whom A talked about the same topic. Without loss of generality, let it be topic T_1.

If any two of these 3 students talked about the topic T_1, then together with the student A they will form a triple that talked about the same topic among themselves. We obtained a contradiction.

If no two of these 3 students talked about the topic T_1, then all 3 students should have talked about the topic T_2, and therefore they form a triple that talked about the same topic among themselves. We obtained a contradiction.

Problem 5 (Combinatorics)

3 squares of size 2×2 are cut off from the board 5×5. Is it always possible to cut off one more 2×2 square?

Solution

Answer: yes, it is always possible.

Let us paint the four corner squares of size 2×2 in black (see Figure 4.2).

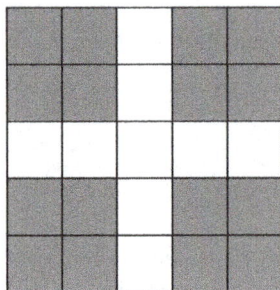

Figure 4.2   Four corner squares of size 2×2 are painted in black in Problem 5.

Notice that when we cut a 2×2 square it can only affect at most one of the 2×2 black squares. Therefore, there exists a black 2×2 square that will not be affected and we can cut it off.

Problem 6 (Number Theory)

Show that the number

$$N = 1^1 + 2^2 + \ldots + 2019^{2019} + 2020^{2020}$$

is divisible by 3.

Solution

We will work modulo 3.

Start by noticing that

$$(6k)^{6k} \equiv (0)^{6k} \equiv 0 \pmod{3}$$
$$(6k+1)^{6k+1} \equiv (1)^{6k+1} \equiv 1 \pmod{3}$$
$$(6k+2)^{6k+2} \equiv (-1)^{6k+2} \equiv 1 \pmod{3}$$
$$(6k+3)^{6k+3} \equiv (0)^{6k+3} \equiv 0 \pmod{3}$$
$$(6k+4)^{6k+4} \equiv (1)^{6k+4} \equiv 1 \pmod{3}$$
$$(6k+5)^{6k+5} \equiv (-1)^{6k+5} \equiv -1 \pmod{3}$$

From here we see that every block of 6 consecutive terms is congruent to

$$0 + 1 + 1 + 0 + 1 + (-1) \equiv 2 \pmod{3}$$

and thus the sum of 18 consecutive terms is divisible by 3.

Since
$$2020 = 2016 + 4 = 18 \cdot 112 + 4$$

then we have that

$$N \equiv 2017^{2017} + 2018^{2018} + 2019^{2019} + 2020^{2020}$$
$$\equiv 1 + 1 + 0 + 1$$
$$\equiv 0 \pmod{3}$$

and the number N is divisible by 3, as desired.

CHAPTER 5

JUNIOR LEVEL EXAM OF 2021

Problem 1 (Combinatorics)

Given an empty table 3×4. In each entry of the table Nirvana writes one of the numbers 0 or 1. After the table is completely filled she calculates the sum of the numbers in each row and the sum of the numbers in each column and writes the obtained results on the board. Prove that there will always be two numbers on the board that are the same.

Solution

The numbers on the board may only take 5 different values:

$$\{0, 1, 2, 3, 4\}$$

However, since there are 3 rows and 4 columns, then there is a total of 7 numbers written on the board. Therefore, by the Pigeonhole Principle there exist two numbers on the board that are the same.

Problem 2 (Geometry)

Point P is chosen inside the triangle ABC, such that $\angle PAB = \angle PBA = 20°$ and $\angle PAC = \angle PCA = 10°$. Prove that the triangle PBC is equilateral.

Solution

The solution presented below refers to Figure 5.1.

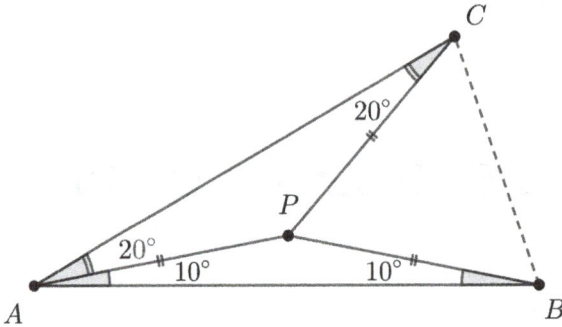

Figure 5.1   Triangle PBC is equilateral in Problem 2.

Since $\angle PAB = \angle PBA$, then the triangle APB is isosceles and $PA = PB$.

Also we have that

$$\angle APB = 180° - \angle PAB - \angle PBA = 160°$$

Since $\angle PAC = \angle PCA$, then the triangle APC is isosceles and $PA = PC$. However, we also have that $PB = PA = PC$, from where the triangle PBC is isosceles.

We also have that

$$\angle APC = 180° - \angle PAC - \angle PCA = 140°$$

From here

$$\angle BPC = 180° - \angle APB - \angle APC = 60°$$

Since we also have that $PB = PC$, then the triangle PBC is equilateral.

Problem 3 (Number Theory)

Find all positive integers n, such that

$$S(n) + n = 11$$

where $S(n)$ is the sum of digits of the number n.

Solution

Answer: $n = 10$.

Notice that since n and $S(n)$ are positive integers, then $n \leq 10$. For $n \leq 9$ we have $S(n) = n$ and therefore

$$S(n) + n = 11$$
$$n + n = 11$$
$$2n = 11$$

which leads to a contradiction.

The number $n = 10$ satisfies the conditions of the problem:

$$S(10) + 10 = 1 + 10 = 11$$

Problem 4 (Combinatorics)

Ten people are sitting around the table. Some of them are *truthers* and always tell the truth while some of them are *liars* and always lie. Some two of them said: "my both neighbors are liars" and other eight of them said: "my both neighbors are truthers". Is it possible that there are more than two truthers?

Solution

Answer: no.

It is obvious that all people cannot be truthers and all people cannot be liars, since then none would be able to say the first phrase.

Let us assume that the number of truthers is more than two. If some two of the truthers are neighbors, then we can go around the table until we reach one of the distributions: truther-truther-liar or liar-truther-truther. However, the truther in the middle would not be able to say any of the phrases. Therefore, all truthers are separated by some of the liars. This implies that each truther should have said the first phrase and then their number is at most two.

We obtained a contradiction.

Problem 5 (Algebra)

Real numbers a, b, c satisfy the equations

$$a + b - c = 0$$
$$a^2 + b^2 - c^2 = 0$$

Prove that they also satisfy the equation

$$a^{2021} + b^{2021} - c^{2021} = 0$$

Solution

Let us rewrite the equations as

$$a + b = c$$
$$a^2 + b^2 = c^2$$

Now let us square both sides of the first equation and substitute c^2 from the second equation:

$$(a + b)^2 = c^2$$
$$a^2 + 2ab + b^2 = c^2$$
$$a^2 + 2ab + b^2 = a^2 + b^2$$
$$\cancel{a^2} + 2ab + \cancel{b^2} = \cancel{a^2} + \cancel{b^2}$$
$$2ab = 0$$
$$ab = 0$$

The last equality implies that either $a = 0$ or $b = 0$. Without loss of generality we can assume that $a = 0$. Then, from the first equation we obtain that $b = c$. Consequently, we have

$$a^{2021} + b^{2021} - c^{2021} = (0)^{2021} + c^{2021} - c^{2021} = 0$$

as desired.

Problem 6 (Number Theory)

Find all positive integers n, such that 4 divides the expression:

$$2^n + n^2 + 1$$

Solution

Answer: $n = 1$.

We will solve the problem by doing the following casework:

- If $n = 1$, then $2^1 + 1^2 + 1 = 4$ is divisible by 4.

- If n is even, then 2^n is even and n^2 is even and, therefore, $2^n + n^2 + 1$ is odd and cannot be divisible by 4.

- If n is odd and $n \geq 3$, then let us put $n = 2k + 1$. Notice that 2^n is divisible by 4. Also

$$n^2 = (2k + 1)^2 = 4k^2 + 4k + 1 \equiv 1 \pmod{4}$$

Therefore
$$2^n + n^2 + 1 \equiv 2 \pmod{4}$$

and cannot be divisible by 4.

We conclude that $2^n + n^2 + 1$ is divisible by 4 only for $n = 1$.

CHAPTER 6

JUNIOR LEVEL EXAM OF 2022

Problem 1 (Combinatorics)

There are 50 people in some class. Is it true that that among them there are at least five who were born in the same month?

Solution

Answer: yes, it is true.

Let us assume that among the people in the class there are no more than 4 born in the same month. Then the total number of people is not greater than

$$4 \cdot 12 = 48$$

which in turn is less than 50. We obtained a contradiction.

Problem 2 (Algebra)

Real numbers a, b, c satisfy the equations

$$a - b = 1$$
$$b - c = 1$$

Junior, Intermediate and Senior Math Olympiads
by Roman Kvasov, Ph.D.

41

Prove that
$$b^2 - ac = 1$$

Solution

Let us rewrite the equations as follows:
$$a = b + 1$$
$$c = b - 1$$

Now let us consider the expression $b^2 - ac$ and substitute a and c:
$$b^2 - ac = b^2 - (b+1)(b-1) = b^2 - b^2 - b + b + 1 = 1$$

as desired.

Problem 3 (Geometry)

In the triangle ABC it is given that $\angle BAC = 60°$. Let AD and BE be the angle bisector and the altitude of the triangle ABC drawn from the vertices A and B respectively. The lines AD and BE intersect at the point P. Prove that $PA = PB$.

Solution

The solution presented below refers to Figure 6.1.

Start by noticing that $\angle PAE = 30°$.

From the triangle APE we have that $\angle APE = 60°$. From here $\angle APB = 120°$.

From the triangle APB we have that $\angle ABP = 30°$. This implies that the triangle APB is isosceles and $PA = PB$.

Problem 4 (Number Theory)

Let $S(n)$ be the sum of digits of the number n. What is the smallest natural number n, such that $S(n)$ and $S(n+1)$ are both even?

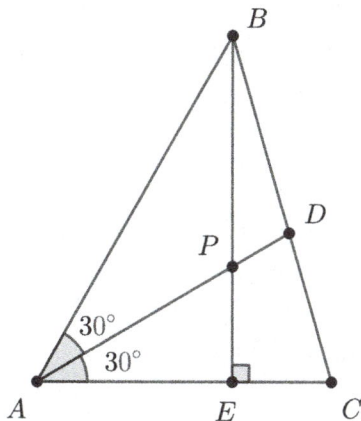

Figure 6.1   Triangle ABP is isosceles in Problem 3.

Solution

Answer: $n = 19$.

Since $S(9) = 9$, then $n \neq 9$. It is not hard to see that $S(i+1)$ and $S(i)$ differ by 1 for $1 \leq i \leq 8$ and $10 \leq i \leq 18$. From here $n \geq 19$.

The number 19 satisfies the conditions of the problem, since $S(19) = 10$ and $S(20) = 2$ are both even numbers.

Problem 5 (Number Theory)

Find all prime numbers p, such that $p^2 + 50$ is also a prime.

Solution

Answer: $p = 3$.

Notice that $p = 3$ satisfies the conditions of the problem, since $(3)^2 + 50 = 59$ is a prime.

If $p \neq 3$, then p is not divisible by 3 and gives remainder 1 or 2 when divided by 3. We will proceed by doing the following casework:

- If $p \equiv 1 \pmod 3$, then

$$p^2 + 50 \equiv (1)^2 + 2 \equiv 0 \pmod 3$$

From here $p^2 + 50$ is divisible by 3, is larger than 3 and, therefore, cannot be a prime.

- If $p \equiv 2 \pmod 3$, then

$$p^2 + 50 \equiv (2)^2 + 2 \equiv 0 \pmod 3$$

From here $p^2 + 50$ is divisible by 3, is larger than 3 and, therefore, cannot be a prime.

We conclude that the only such value is $p = 3$.

Problem 6 (Combinatorics)

In the beginning there is a box with 20 danish cookies. Mara and Arturo are playing the following game. Mara goes first and eats no more than half of the cookies present in the box. Then goes Arturo and eats no more than half of the cookies that are left in the box, etc. At each turn the players can only eat no more than half of the cookies left in the box until one of them cannot eat any cookies and loses the game. Who has a winning strategy?

Solution

Answer: Mara has a winning strategy.

Working backwards we can find the "winning positions" for Mara in this game to be 1, 4, 7 and 15.

If Mara leaves Arturo 1 cookie, then he wins. If Mara leaves Arturo 2 cookies, then he loses.

If Mara leaves Arturo 3 cookies, then he wins. It is not hard to see that if Mara leaves Arturo 4, 5 or 6 cookies, then he will loose, as Arturo will be able to make a move and leave Mara 3 cookies.

If Mara leaves Arturo with 7 cookies, then he wins. Similarly, if Mara leaves Arturo any of the 8, 9, ... , 14 cookies, then he will loose and if Mara leaves Arturo 15 cookies, then he will win.

Therefore, Mara's first move should be to eat 5 cookies, leaving Arturo 15 cookies and then 7, 3 and 1 cookie.

CHAPTER 7

JUNIOR LEVEL EXAM OF 2023

Problem 1 (Combinatorics)

Integer numbers from 4 to 14, inclusive, are arranged in a row. Can the symbols "$+$" and "$-$" be inserted between them in such a way that the value of the resulting expression is zero?

Solution

Answer: no.

Start by noticing that among the written numbers there are exactly 5 odd numbers:
$$5, 7, 9, 11, 13$$
Also, notice that adding or subtracting an odd number changes the parity of the result. Since there is an odd number of odd numbers, then the resulting expression will always be odd and cannot be equal to zero.

Problem 2 (Number Theory)

Let $S(n)$ be the sum of digits of the number n. What is the smallest positive integers n, such that
$$S\left(n^2\right) - S(n) = 3$$

Junior, Intermediate and Senior Math Olympiads
by Roman Kvasov, Ph.D.

45

Solution

Answer: $n = 4$.

Notice that $n = 1, 2, 3$ do not work, since

$$S(1) - S(1) = 0$$
$$S(4) - S(2) = 2$$
$$S(9) - S(3) = 6$$

The number $n = 4$ satisfies the conditions of the problem:

$$S(16) - S(4) = 3$$

as desired.

Problem 3 (Combinatorics)

There are two piles of stones. The first pile has 2022 stones and the second pile has 2023 stones. Two boys are playing a game, where at each turn, a boy may take as many stones as he likes, but only from one of the piles. The loser is the player who cannot make a move. Who will win the game?

Solution

Answer: the first player has a winning strategy.

The first player with his first move should take 1 stone from the pile that has 2023 stones. With every next move he should respond "symmetrically" to the moves of the second player: if the second player take x stones from some pile, then the first player should take x stones from the other pile. This way the first player will always keep the numbers of the stones equal in both piles and will be the last one to be able to make a move.

Problem 4 (Algebra)

Positive real numbers a, b and c satisfy the equations

$$a^2 - b = c$$
$$b^2 - a = c$$

Is it necessary that $a = b$?

Solution

Answer: yes.

Let us subtract the given equations and factor the left-hand side using the Difference of Squares Formula:

$$\left(a^2 - b\right) - \left(b^2 - a\right) = 0$$
$$a^2 - b - b^2 + a = 0$$
$$(a - b)(a + b) + (a - b) = 0$$
$$(a - b)(a + b + 1) = 0$$

The last equation implies that $a - b = 0$ or $a + b + 1 = 0$. Since the numbers a and b are positive, then $a + b + 1$ is also positive, and, therefore, cannot be equal to zero. From here we have that $a - b = 0$, or equivalently $a = b$, as desired.

Problem 5 (Number Theory)

Let m and n be integer numbers, such that $m^2 + 2n^2$ is divisible by 5. Prove that $2m^2 + n^2$ is also divisible by 5.

Solution

Let us consider the following table of residues modulo 5:

$x \pmod 5$	$x^2 \pmod 5$
0	$(0)^2 \equiv 0$
1	$(1)^2 \equiv 1$
2	$(2)^2 \equiv 4 \equiv -1$
3	$(3)^2 \equiv 9 \equiv -1$
4	$(4)^2 \equiv 16 \equiv 1$

From here we see that a number that is a square of an integer is congruent to 0, 1 or -1 modulo 5. This implies that a number that is twice a square of an integer is congruent to 0, 2 or -2 modulo 5. Therefore, the number $m^2 + 2n^2$ is divisible by 5 if and only if both numbers m and n are divisible by 5. From here

$$2m^2 + n^2 \equiv 2(0)^2 + (0)^2 \equiv 0 \pmod{5}$$

and $2m^2 + n^2$ is also divisible by 5, as desired.

Problem 6 (Geometry)

Given a square $ABCD$ and the point E that lies outside of the square, such that $\angle ACE = 105°$ and

$$AD + DE = DC + CE$$

Prove that $\angle BAE = 75°$.

Solution

The solution presented below refers to Figure 7.1.

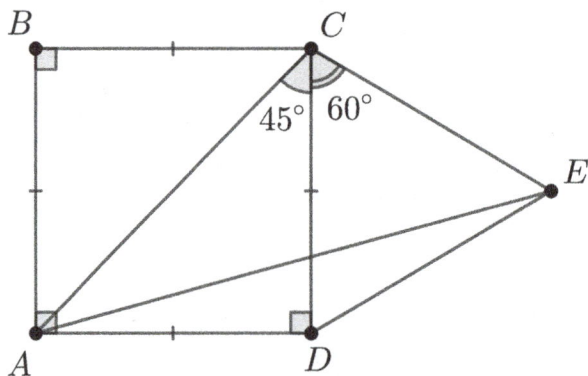

Figure 7.1   Triangle ADE is isosceles in Problem 6.

Notice that since $\angle ACD = 45°$, then

$$\angle DCE = \angle ACE - \angle ACD = 105° - 45° = 60°$$

Since $ABCD$ is a square, then $AD = DC$, and, therefore $DE = CE$. From here the triangle CDE is equilateral and $CD = DE = EC$.

Since $AD = DC = DE$, then the triangle ADE is isosceles and

$$\angle DAE = \frac{1}{2}\left(180° - \angle ADE\right) = \frac{1}{2}\left(180° - 150°\right) = 15°$$

From here

$$\angle BAE = 90° - 15° = 75°$$

as desired.

CHAPTER 8

JUNIOR LEVEL EXAM OF 2024

Problem 1 (Number Theory)

Let $S(n)$ be the sum of digits of the number n. What is the largest three-digit integer n, such that

$$S(n) < 3$$

Solution

Answer: $n = 200$.

Notice that either $S(n) = 1$ or $S(n) = 2$.

If $S(n) = 1$, then $n = 100$.

If $S(n) = 2$, then n is equal to 101, 110 or 200, and the largest three-digit integer n is 200.

Problem 2 (Geometry)

Given a right triangle ABC, where $\angle ABC = 90°$ and $\angle BAC = 30°$. Let CD be the angle bisector of the angle $\angle ACB$. Prove that $AD = DC$.

Junior, Intermediate and Senior Math Olympiads
by Roman Kvasov, Ph.D.

51

Solution

The solution presented below refers to Figure 8.1.

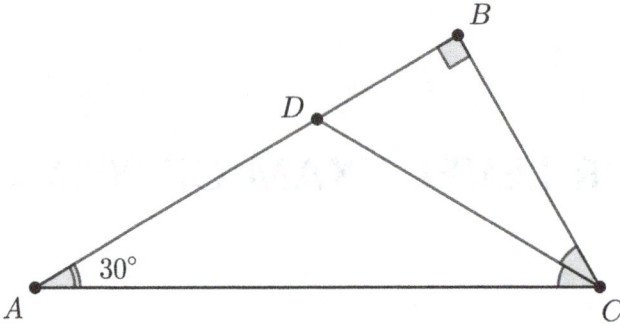

Figure 8.1   Triangle ACD is isosceles in Problem 2.

From the triangle ABC we have

$$\angle ACB = 180° - \angle ABC - \angle BAC = 180° - 90° - 30° = 60°$$

Since CD is the angle bisector of the angle $\angle ACB$, then

$$\angle BCD = \angle ACD = 30°$$

This implies that the triangle ACD is isosceles and $AD = DC$, as desired.

Problem 3 (Algebra)

Distinct real numbers a and b satisfy the equations

$$a^2 + c^2 = a$$
$$b^2 + c^2 = b$$

Find the value of $a + b$.

Solution

Answer: $a + b = 1$.

We will subtract the given equations and factor the left-hand side using the Difference of Squares Formula:

$$a^2 - b^2 = a - b$$
$$(a - b)(a + b) - (a - b) = 0$$
$$(a - b)(a + b - 1) = 0$$

The last equation implies that $a - b = 0$ or $a + b - 1 = 0$. Since the numbers a and b are distinct, then $a + b - 1 = 0$ or equivalently $a + b = 1$, as desired.

Problem 4 (Combinatorics)

There are three piles of stones: the first has 10, the second has 15, and the third has 20. On each turn, it's allowed to split any pile into two smaller ones. The player who cannot make a move loses. Who will win?

Solution

Answer: second player will win.

Start by noticing that after each move, the number of piles increases by 1. Initially, there were 3 piles, and in the end, there are 45. Thus, a total of 42 moves will be made. The winner will make the winning move on the 42nd turn, and it will be the second player.

Problem 5 (Geometry)

In the triangle ABC the squares $ABPM$ and $CBQN$ are built outwardly. Prove that $AQ = CP$.

Solution

The solution presented below refers to Figure 8.2.

Start by noticing that $AB = BP$ and $BC = BQ$. Also

$$\angle ABQ = \angle ABC + 90° = \angle PBC$$

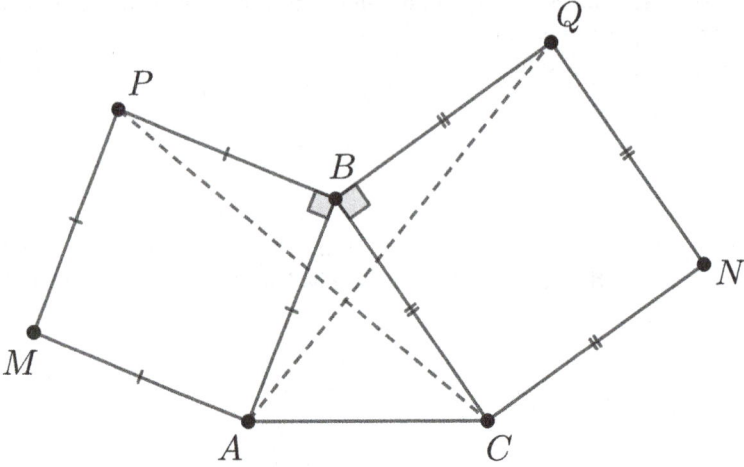

Figure 8.2   Triangles ABQ and PBC are congruent in Problem 5.

From here the triangles ABQ and PBC are congruent by Side-Angle-Side Postulate. From here $AQ = CP$, as desired.

Problem 6 (Number Theory)

Solve the equation in integer numbers

$$m^{2222} - 2222 = 7n$$

Solution

Answer: the equation has no integer solutions.

Let us assume that there exist integer solutions to the given equation and let us rewrite it as follows:

$$k^2 = 7n + 2222$$

where $k = m^{1111}$.

Notice that 2222 gives remainder 3 when divided by 7. Therefore, the equation implies that a square of an integer gives remainder 3 when divided by 7. Let us consider the following table of residues modulo 7:

$m \pmod 7$	$m^2 \pmod 7$
0	$(0)^2 \equiv 0$
1	$(1)^2 \equiv 1$
2	$(2)^2 \equiv 4$
3	$(3)^2 \equiv 9 \equiv 2$
4	$(4)^2 \equiv 16 \equiv 2$
5	$(5)^2 \equiv 25 \equiv 4$
6	$(6)^2 \equiv 36 \equiv 1$

From here we see that a square of an integer is always congruent to either 0, 1, 2 or 4 modulo 7. We obtained a contradiction, and, therefore, there are no integer solutions to this equation.

CHAPTER 9

TOPICS FOR JUNIOR MATH OLYMPIADS

Operations on Fractions

For a, b, c, d and n being real numbers, the following holds:

1. $\frac{a}{b} \cdot \frac{c}{d} = \frac{ac}{bd}$

2. $\frac{a}{b} \cdot \frac{b}{c} = \frac{a}{c}$

3. $\frac{a}{b} \div \frac{c}{d} = \frac{ad}{bc}$

4. $\frac{\frac{a}{b}}{\frac{c}{d}} = \frac{ad}{bc}$

5. $\frac{a}{c} + \frac{b}{c} = \frac{a+b}{c}$

6. $\frac{a}{c} - \frac{b}{c} = \frac{a-b}{c}$

7. $\frac{a}{b} + \frac{c}{d} = \frac{ad+bc}{bd}$

8. $\frac{a}{b} - \frac{c}{d} = \frac{ad-bc}{bd}$

9. $\frac{a}{a} = 1$

10. $\left(\frac{a}{b}\right)^{-1} = \frac{b}{a}$

11. $\left(\frac{a}{b}\right)^{n} = \frac{a^n}{b^n}$

12. $\left(\frac{a}{b}\right)^{-n} = \frac{b^n}{a^n}$

Junior, Intermediate and Senior Math Olympiads
by Roman Kvasov, Ph.D.

57

Basic Factorization Formulas

For real numbers a and b the following formulas hold:

1. $a^2 - b^2 = (a - b)(a + b)$

2. $a^3 + b^3 = (a + b)\left(a^2 - ab + b^2\right)$

3. $a^3 - b^3 = (a - b)\left(a^2 + ab + b^2\right)$

4. $a^2 + 2ab + b^2 = (a + b)^2$

5. $a^2 - 2ab + b^2 = (a - b)^2$

6. $a^3 + 3a^2b + 3ab^2 + b^3 = (a + b)^3$

7. $a^3 - 3a^2b + 3ab^2 - b^3 = (a - b)^3$

Properties of Exponents

Let $a > 0$, then

1. $a^m \cdot a^n = a^{m+n}$

2. $\dfrac{a^m}{a^n} = a^{m-n}$

3. $(a^m)^n = a^{m \cdot n}$

4. $(a \cdot b)^n = a^n \cdot b^n$

5. $\left(\dfrac{a}{b}\right)^n = \dfrac{a^n}{b^n}$

Common Factor Factorization

The common factor factorization technique is based on the inverted distributive properties:

1. $ab + ac = a(b + c)$

2. $ab - ac = a(b - c)$

Factorization by Grouping

This factorization technique is based on repeatedly applying common factor factorization. In the following example we factor out the common factors a, b and $x + y$ consecutively:

$$ax + ay + bx + by = (ax + ay) + (bx + by)$$
$$= a(x + y) + b(x + y)$$
$$= (x + y)(a + b)$$

Simon's Favorite Factoring Trick

This factorization technique is based on grouping and allows for partial factoring of certain expressions. For example

$$xy + 2x + 3y = (xy + 2x) + (3y + 6) - 6$$
$$= x(y + 2) + 3(y + 2) - 6$$
$$= (y + 2)(x + 3) - 6$$

Decimal Representation

Every positive integer n can be expressed uniquely in the following form:

$$n = a_m \cdot 10^m + a_{m-1} \cdot 10^{m-1} + \ldots + a_1 \cdot 10 + a_0$$

where a_i are digits and $a_m \neq 0$. This is usually written as

$$n = \overline{a_m a_{m-1} \ldots a_1 a_0}$$

Divisibility by a Number

The integer number m is divisible by a nonzero integer number n if there exists an integer number k, such that

$$m = n \cdot k$$

The notation used for divisibility is $n \mid m$, which is read as "n divides m".

Properties of Divisibility

1. For all nonzero integers a:
    $$a \mid a$$

2. For integers a, b and c if $a \mid b$ and $b \mid c$, then
    $$a \mid c$$

3. For integers a, b and c if $a \mid b$, then
    $$a \mid b \cdot c$$

4. For integers a, b and positive integer k if $a \mid b$, then
    $$a \mid b^k$$

5. For integers a, b and c if $a \mid b$ and $a \mid c$, then
    $$a \mid b + c$$

6. For integers a, b and c if $a \mid b$ and $a \mid c$, then
    $$a \mid b - c$$

7. For integers a, b and c if $a \mid b$ and $a \mid b \pm c$, then
    $$a \mid c$$

8. For integer a, b and m if $am \mid bm$, then
    $$a \mid b$$

9. For integers a and b if $a \mid b$ and $b \mid a$, then
    $$|a| = |b|$$

10. For integers a and b if $a \mid b$ and $b \neq 0$, then
    $$|a| \leq |b|$$

Remainder by a Modulo

Let n be a positive integer number that we will call **modulo**. If we divide an integer m by n, the result of the division is k and the remainder is r, such that

$$0 \leq r \leq n - 1$$

The number n can be written in the following form:

$$m = k \cdot n + r$$

Composite and Prime Numbers

A positive integer number $n > 1$ is called **composite** if it has a positive integer divisor apart from 1 and itself. If the only divisors of the integer $n > 1$ are 1 and itself, then it is called **prime**.

Fundamental Theorem of Arithmetic

Every positive integer greater than 1 can be represented uniquely as a product of prime numbers, up to the order of the factors.

Divisibility Rules

1. **Divisibility Rule for 2**. A number is divisible by 2 if its last digit is divisible by 2.

2. **Divisibility Rule for 3**. A number is divisible by 3 if the sum of its digits is divisible by 3.

3. **Divisibility Rule for 4**. A number is divisible by 4 if the number formed by its last two digits is divisible by 4.

4. **Divisibility Rule for 5**. A number is divisible by 5 if its last digit is 0 or 5.

5. **Divisibility Rule for 6**. A number is divisible by 12 if it satisfies the divisibility rules for both 2 and 3.

6. **Divisibility Rule for 8**. A number is divisible by 8 if the number formed by its last three digits is divisible by 8.

7. **Divisibility Rule for 9**. A number is divisible by 9 if the sum of its digits is divisible by 9.

8. **Divisibility Rule for 10**. A number is divisible by 10 if its last digit is 0.

9. **Divisibility Rule for 11**. A number is divisible by 11 if the difference between the sum of its digits at even places and the sum of its digits at odd places is divisible by 11.

10. **Divisibility Rule for 12**. A number is divisible by 12 if it satisfies the divisibility rules for both 3 and 4.

Congruence

The integer numbers a and b are congruent by a modulo n if $n|(a - b)$. The notation used for congruence is

$$a \equiv b \pmod{n}$$

Properties of Congruence

For positive integers a, b, c, d, e, such that that

$$a \equiv b \pmod{n}$$

and

$$c \equiv d \pmod{n}$$

The following holds:

1. $a + e \equiv b + e \pmod{n}$

2. $a - e \equiv b - e \pmod{n}$

3. $a \cdot e \equiv b \cdot e \pmod{n}$

4. $a^e \equiv b^e \pmod{n}$ where e is a positive integer

5. $a + c \equiv b + d \pmod{n}$

6. $a - c \equiv b - d \pmod{n}$

7. $a \cdot c \equiv b \cdot d \pmod{n}$

Pigeonhole Principle

Pigeonhole Principle states that if n items are put into m containers, with $n > m$, then at least one container must contain more than one item.

Sum of Angles in a Triangle

The sum of the angles in any triangle is $180°$

Angles and Parallel Lines

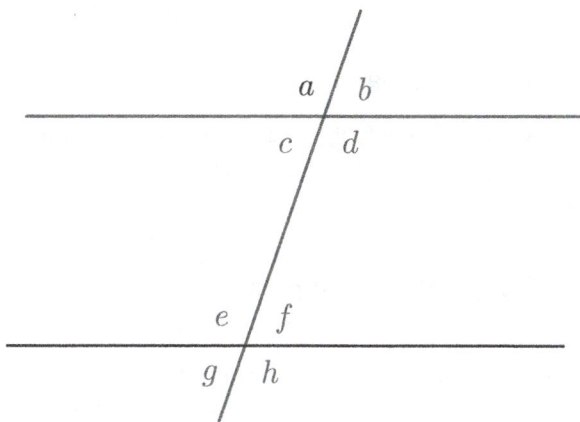

Let two parallel lines be intersected by a transversal, such that 8 angles are formed: a, b, c, d, e, f, g and h.

The definitions of the formed angles are given below:

1. pairs of angles (a, d), (b, c), (e, h), (f, g) are called **vertical angles**

2. pairs of angles (a, b), (c, d), (e, f), (g, h) are called **linear pair angles**

3. pairs of angles (a, e), (b, f), (c, g), (d, h) are called **corresponding angles**

4. pairs of angles (c, f), (d, e) are called **alternate interior angles**

5. pairs of angles (a, h), (b, g) are called **alternate exterior angles**

6. pairs of angles (c, e), (d, f) are called **consecutive interior angles**

The properties of the formed angles are given below:

1.   vertical angles are congruent

2.   linear pair angles add up to $180°$

3.   corresponding angles are congruent

4.   alternate interior angles are congruent

5.   alternate exterior angles are congruent

6.   consecutive interior angles add up to $180°$

Isosceles Triangles

- Triangle is *isosceles* if and only if two of its sides are congruent.

- Triangle is *isosceles* if and only if two of its angles are congruent.

Postulates for Congruent Triangles

- *Side-Angle-Side Postulate* states that if two sides and the included angle in one triangle are congruent to two sides and the included angle in another triangle, then the two triangles are congruent.

- *Angle-Side-Angle Postulate* states that if two angles and one side in one triangle are congruent to the corresponding two angles and one side in another triangle, then the two triangles are congruent.

- *Side-Side-Side Postulate* states if three sides in one triangle are congruent to three sides in another triangle, then the triangles are congruent.

- *Hypotenuse-Leg Postulate* states if the hypotenuse and leg in one right triangle are congruent to the hypotenuse and leg in another right triangle, then the triangles are congruent.

PART II

INTERMEDIATE MATH OLYMPIADS

Intermediate Level math olympiads are tailored for students up to the 10th grade, presenting a valuable opportunity for participants with some exposure to olympiad mathematics and proof writing. These competitions encompass a diverse array of fundamental and advanced topics in algebra, geometry, combinatorics, and number theory, aiming to challenge and deepen the understanding of mathematical concepts among young minds.

In the realm of algebra, participants can anticipate encountering problems involving the Inequality of Arithmetic and Geometric Means, Titu's Lemma, Cauchy-Schwarz inequality, systems of equations, various factorization techniques, Vieta's formulas, sequences, and the adept use of formulas, along with proficiency in tackling basic functional equations. These algebraic principles function as foundational elements for more advanced mathematical reasoning.

Geometry problems at the Intermediate level often revolve around angle chase, cyclic quadrilaterals, inscribed and central angles, transformations, and the properties of the orthocenter, centroid, incenter, and circumcenter. The emphasis is on refining advanced geometric principles and improving problem-solving skills.

Combinatorics introduces participants to concepts such as Induction, the Pigeonhole principle, invariants and monovariants, more intricate coloring techniques, and different strategies for games. These problems not only assess mathematical reasoning but also foster creative thinking and logical deduction.

In the area of number theory, Intermediate-level participants explore topics such as greatest common divisor and least common multiple, Diophantine equations, modular arithmetic, Euclidean algorithm, Chinese remainder theorem, employing estimation techniques and inequalities. These problems delve into the fascinating world of number theory, engaging participants in the exploration of patterns and relationships within the realm of integers.

To prepare effectively for Intermediate Math Olympiads, aspiring participants should concentrate on mastering these fundamental concepts. This can be achieved through a combination of regular practice, exposure to a variety of problems, and participation in mock exams. Developing a profound understanding of the basic principles in algebra, geometry, combinatorics, and number theory will empower students to approach olympiad problems with confidence and creativity. Additionally, seeking guidance from experienced mentors or teachers and collaborating with peers in problem-solving sessions can further enhance one's preparation for the challenges presented by Intermediate math olympiads.

CHAPTER 1

INTERMEDIATE LEVEL EXAM OF 2017

Problem 1 (Number Theory)

Prove that the number
$$N = 7^{2017} + 2017^7$$
is divisible by 8.

Solution

Working modulo 8 we have
$$N = 7^{2017} + 2017^7 \equiv (-1)^{2017} + (1)^7 \equiv -1 + 1 \equiv 0 \pmod 8$$
which implies that N is divisible by 8, as desired.

Problem 2 (Algebra)

Find all functions $f : \mathbb{R} \to \mathbb{R}$ that satisfy the equation
$$f(x + y) - f(x - y) = f(x) - f(y)$$
for all real numbers x and y.

Junior, Intermediate and Senior Math Olympiads
by Roman Kvasov, Ph.D.

67

Solution

Answer: $f(x) = c$, where c is a real constant.

It is not hard to check that all constant functions satisfy this functional equation.

Let us put $f(0) = c$ and substitute $y = 0$ into the functional equation:

$$f(x + 0) - f(x - 0) = f(x) - f(0)$$
$$f(x) - f(x) = f(x) - c$$
$$0 = f(x) - c$$
$$c = f(x)$$

and consequently, $f(x) = c$, as desired.

Problem 3 (Geometry)

Let ABC be a triangle with $AB > AC$. The angle bisectors at B and C meet at point I inside the triangle ABC. The circumcircle of the triangle BIC intersects AB again in X and AC again in Y. Show that CX is parallel to BY.

Solution

The solution presented below refers to Figure 1.1.

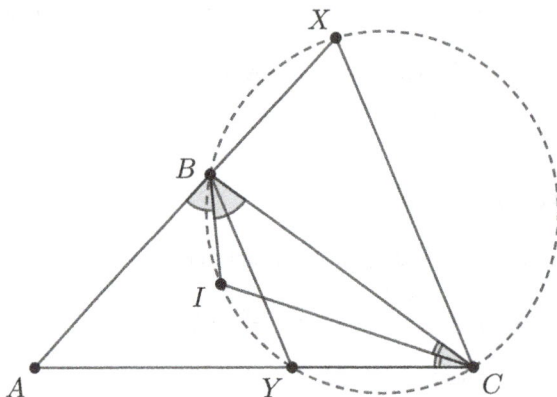

Figure 1.1 Quadrilaterals $BICY$ and $XBCI$ are cyclic in Problem 3.

Let $\angle IBC = \beta$ and $\angle ICB = \gamma$. Then we have

$$\angle BIC = 180° - \beta - \gamma$$

Since $BICY$ is cyclic, then

$$\angle BYC = 180° - \angle BIC = \beta + \gamma$$

Since $XBCI$ is cyclic, then

$$\angle XBI = \angle XCI = \beta$$

and therefore

$$\angle XCQ = \beta + \gamma$$

and $CX \parallel BY$, as desired.

Problem 4 (Number Theory)

Find the greatest common divisor d of the set of numbers

$$2017 + 1, 2017^2 + 1, \ldots, 2017^{2017} + 1$$

Solution

Answer: $d = 2$.

Start by noticing that all numbers in the given set are even. Therefore, d is also an even number.

Notice that d and 2017 are relatively prime. Since d divides $(2017 + 1)$ and $(2017^2 + 1)$, then d divides their difference:

$$(2017^2 + 1) - (2017 + 1) = 2017^2 - 2017 = 2017 \cdot 2016$$

Since d is relatively prime with 2017, then d divides 2016.

Since d divides $(2017 + 1)$, then d should also divide the difference:

$$(2017 + 1) - 2016 = 2$$

Since d divides 2 and d is an even number, then $d = 2$, as desired.

Problem 5 (Combinatorics)

Given an empty table $3 \times n$, where n is a positive integer. In each entry of the table Carlos writes one of the numbers 0 or 1. After the table is completely filled he calculates the sum of the numbers in each row and the sum of the numbers in each column and writes the obtained results on the board. Find the largest value of n, such that among the numbers on the board no three are the same.

Solution

Answer: $n = 7$.

Notice that there are only 4 possible different sums in columns: 0, 1, 2 or 3.

Let us assume that $n \geq 9$. Since

$$9 > 2 \cdot 4$$

then by the Pigeonhole Principle we have that there will be at least three equal sums among the sums in columns.

Let us assume that for $n = 8$ among the numbers on the board no three are the same. Therefore, in the 8 columns the 4 possible different sums should repeat exactly once each. The sum of all numbers in the table is

$$0 + 0 + 1 + 1 + 2 + 2 + 3 + 3 = 12$$

If the sums in the rows are all equal to 4, then we have a contradiction. Otherwise the smallest of them is less than 4 and will repeat three times with the corresponding sum in the row and we have a contradiction.

For $n = 7$, an example of the table where no three of the sums are the same is given below:

0	0	0	0	0	1	1
0	0	1	0	1	1	1
0	0	0	1	1	1	1

Problem 6 (Algebra)

Prove the inequality for positive values of a, b, c, such that $a + b + c = 1$:

$$\frac{ab}{b^2 + b + 1} + \frac{bc}{c^2 + c + 1} + \frac{ca}{a^2 + a + 1} \leq \frac{1}{3}$$

Solution

Let us apply the AM-GM Inequality to each of the following denominators:

$$a^2 + a + 1 \geq 3\sqrt[3]{a^3} = 3a$$
$$b^2 + b + 1 \geq 3\sqrt[3]{b^3} = 3b$$
$$c^2 + c + 1 \geq 3\sqrt[3]{c^3} = 3c$$

From here

$$\frac{ab}{b^2+b+1} + \frac{bc}{c^2+c+1} + \frac{ca}{a^2+a+1} \leq \frac{ab}{3b} + \frac{bc}{3c} + \frac{ca}{3a}$$
$$= \frac{a}{3} + \frac{b}{3} + \frac{c}{3}$$
$$= \frac{a+b+c}{3}$$
$$= \frac{1}{3}$$

which is what needed to be proven.

CHAPTER 2

INTERMEDIATE LEVEL EXAM OF 2018

Problem 1 (Number Theory)

Given the integer numbers x and y. It is known that $x + y$ is divisible by 3. Prove that $x^2 - xy + y^2$ is also divisible by 3.

Solution

Since $x + y$ is divisible by 3, then

$$x + y \equiv 0 \pmod{3}$$

or equivalently

$$x \equiv -y \pmod{3}$$

Now we have

$$x^2 - xy + y^2 \equiv (-y)^2 - (-y)y + y^2$$
$$\equiv y^2 + y^2 + y^2$$
$$\equiv 3y^2$$
$$\equiv 0 \pmod{3}$$

and, therefore $x^2 - xy + y^2$ is divisible by 3, as desired.

Junior, Intermediate and Senior Math Olympiads
by Roman Kvasov, Ph.D.

73

Problem 2 (Algebra)

Numbers a and b satisfy the equalities

$$b + 1 = a^2 + \frac{1}{a^2}$$

$$b^2 + 1 = a^3 + \frac{1}{a^3}$$

Prove that

$$b + \frac{1}{b} = a + \frac{1}{a}$$

Solution

Let us consider the second equation, factor its right-hand side and substitute $a^2 + \frac{1}{a^2}$ for $b + 1$:

$$b^2 + 1 = a^3 + \frac{1}{a^3}$$

$$b^2 + 1 = \left(a + \frac{1}{a}\right)\left(a^2 + \frac{1}{a^2} - 1\right)$$

$$b^2 + 1 = \left(a + \frac{1}{a}\right)(b + 1 - 1)$$

$$b^2 + 1 = \left(a + \frac{1}{a}\right)b$$

$$\frac{b^2 + 1}{b} = a + \frac{1}{a}$$

$$\frac{b^2}{b} + \frac{1}{b} = a + \frac{1}{a}$$

$$b + \frac{1}{b} = a + \frac{1}{a}$$

which is what needed to be proven.

Problem 3 (Geometry)

Let M be the midpoint of the side BC of the triangle ABC. Point K is chosen on the segment AM, such that $AK = BM$. It is known that $\angle AMC = 60°$. Prove that $AC = BK$.

Solution

The solution presented below refers to Figure 2.1.

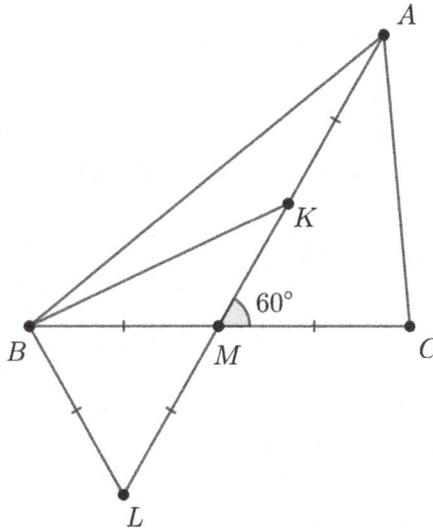

Figure 2.1   Triangles AMC and BLK are congruent Problem 3.

Let us construct the point L on the line AM, such that M lies between K and L and $LM = BM$. Notice that from here the triangle BML is equilateral.

The triangles AMC and BLK are congruent by Side-Angle-Side Postulate. Therefore $BK = AC$, as desired.

Problem 4 (Algebra)

Let a, b, c be positive real numbers such that $a + b + c = 6$. Show that

$$\frac{a^2}{a+b} + \frac{b^2}{b+c} + \frac{c^2}{c+a} \geq 3$$

Solution

We will apply the Titu's Lemma for the numbers (a, b, c) and $(a+b, b+c, c+a)$.

We have

$$\frac{a^2}{a+b} + \frac{b^2}{b+c} + \frac{c^2}{c+a} \geq \frac{(a+b+c)^2}{2(a+b+c)} = \frac{(6)^2}{2(6)} = 3$$

as desired.

Problem 5 (Combinatorics)

Given a board 8×8. You can use any figures of the form:

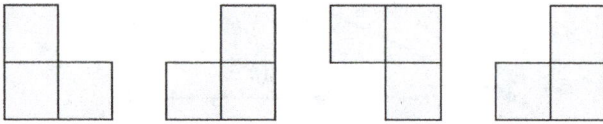

Is it possible to place 10 figures on the board in such a way that one additional figure cannot be placed without overlapping?

Solution

Answer: no.

Notice that 10 figures cover $3 \cdot 10 = 30$ squares. Now let us assume that it is possible to place 10 figures, such that it is impossible to place one more figure without overlapping.

Let us divide the board into 16 squares 2×2 (see Figure 2.2). In each square 2×2 at least two squares should be covered (otherwise we will be able to place a figure). Therefore, at least $16 \cdot 2 = 32$ squares should be covered. Contradiction.

Problem 6 (Number Theory)

Find the smallest natural number n, such that n is divisible by 99 and all its digits are even.

Solution

Answer: 228888.

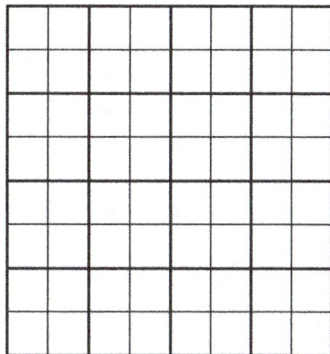

Figure 2.2   Board 8×8 is divided into 16 squares 2×2 in Problem 5.

Let us assume that the number n has an even number of digits (otherwise we will complement it with zero before its' first digit):

$$n = \overline{a_{2k}a_{2k-1}\ldots a_4 a_3 a_2 a_1}$$

Let us define the positive integer number s as follows:

$$s = \overline{a_{2k}a_{k-1}} + \ldots + \overline{a_4 a_3} + \overline{a_2 a_1}$$

Then we have that

$$n = \sum_{i=1}^{k} 100\overline{a_{2i}a_{i-1}} \equiv \sum_{i=1}^{k} \overline{a_{2i}a_{i-1}} \equiv s \pmod{99}$$

Let us show that n has at least 6 digits, i.e. $k \geq 3$. Indeed, since s is even and is divisible by 99, then it is also divisible by 198 and therefore $s \geq 198$. Now, if $k < 3$, then

$$0 < s \leq 8 + 88 + 88 = 184 < 198$$

which leads to a contradiction. Therefore, we have shown that $k \geq 3$.

Let us assume that $k = 3$ and $n = \overline{a_6 a_5 a_4 a_3 a_2 a_1}$. Notice that if $\overline{a_6 a_5} < 22$, then

$$0 < s < 22 + 88 + 88 = 198$$

which also leads to a contradiction. From here $\overline{a_6 a_5}$ is at least 22. It is now enough to notice that the number 228888 satisfies the conditions of the problem.

CHAPTER 3

INTERMEDIATE LEVEL EXAM OF 2019

Problem 1 (Number Theory)

Show that if an integer number n is not divisible by 5, then $n^2 - 1$ or $n^2 + 1$ should be divisible by 5.

Solution

We will work modulo 5.

For integers n not divisible by 5, we will consider the table of residues modulo 5:

$n \pmod 5$	$n^2 \pmod 5$
1	$(1)^2 \equiv 1$
2	$(2)^2 \equiv -1$
3	$(3)^2 \equiv 9 \equiv -1$
4	$(4)^2 \equiv 16 \equiv 1$

From here we see that n^2 is always congruent to either 1, or -1 modulo 5.

If n^2 is congruent to 1 modulo 5, then $n^2 - 1$ is divisible by 5. If n^2 is congruent to -1 modulo 5, then $n^2 + 1$ is divisible by 5. This completes the proof.

Junior, Intermediate and Senior Math Olympiads
by Roman Kvasov, Ph.D.

79

Problem 2 (Algebra)

Given two distinct nonzero real numbers a and b that satisfy the equation

$$a + \frac{1}{a} = b + \frac{1}{b}$$

Prove that one of the numbers is a reciprocal of another.

Solution

Notice that it will enough to prove that $ab = 1$. Let us multiply the equation by ab and rewrite it as follows:

$$a + \frac{1}{a} = b + \frac{1}{b}$$

$$ab \cdot \left(a + \frac{1}{a}\right) = ab \cdot \left(b + \frac{1}{b}\right)$$

$$a^2b + b = ab^2 + a$$

$$a^2b + b - ab^2 - a = 0$$

$$\left(a^2b - ab^2\right) - (a - b) = 0$$

$$ab(a - b) - (a - b) = 0$$

$$(a - b)(ab - 1) = 0$$

Since the numbers are distinct then from the last equation we have that $ab - 1 = 0$, or equivalently $ab = 1$, as desired.

Problem 3 (Number Theory)

Find all positive integers m and n, such that

$$2^{m-1} + 3 = n^2$$

Solution

Answer: $m = 1$, $n = 2$.

We will solve the problem by doing the following casework:

- If $m = 1$, then the equation becomes

$$4 = n^2$$

and consequently, $n = 2$ is the only positive integer in this case.

- If $m = 2$, then the equation becomes

$$5 = n^2$$

and consequently, there are no such positive integer n in this case.

- If $m \geq 3$, then we will work modulo 4. Since 2^{m-1} is divisible by 4, then the right-hand side is congruent to 3 modulo 4. However, no perfect square is congruent to 3 mod 4, and consequently, there are no solutions in this case.

Problem 4 (Geometry)

Given a triangle ABC, such that $\angle ACB = 120°$. Points D and E are chosen on the side AB, such that $DC \perp AC$ and $CE \perp AB$. Given that $AE = EB$, prove that $DB = DC$.

Solution

The solution presented below refers to Figure 3.1.

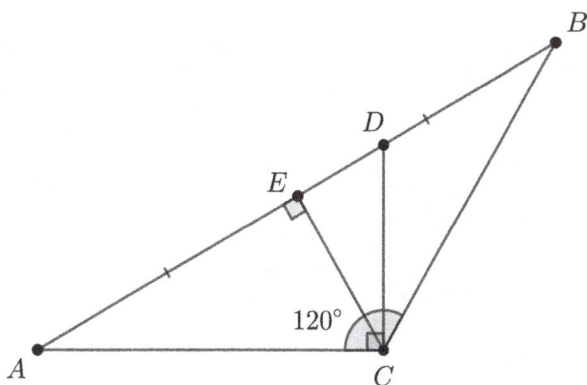

Figure 3.1   Triangles AEC and BEC are congruent Problem 4.

Since $CE \perp AB$ and $AE = EB$, then the triangles AEC and BEC are congruent by Side-Angle-Side Postulate. From here

$$\angle BAC = \angle ABC = 30°$$

Since $DC \perp AC$, then $\angle ACD = 90°$ and

$$\angle DCB = \angle ACB - \angle ACD = 120° - 90° = 30°$$

Therefore, the triangle CDB is isosceles and $DB = DC$, as desired.

Problem 5 (Algebra)

Find all positive integer numbers $n \geq 3$, such that there exist n positive real numbers $x_1, x_2, \ldots, x_n$ that satisfy the equations:

$$x_1 + x_2 + \ldots + x_n = 3$$

$$\frac{1}{x_1} + \frac{1}{x_2} + \ldots + \frac{1}{x_n} = 3$$

Solution

Answer: $n = 3$.

It is not hard to see that $n = 3$ satisfies the conditions of the problem for the numbers $x_1 = 1$, $x_2 = 1$, $x_3 = 1$.

Let us now assume that $n \geq 4$. Let us multiply the equations and apply the AM-GM Inequality as follows:

$$9 = (x_1 + x_2 + \ldots + x_n)\left(\frac{1}{x_1} + \frac{1}{x_2} + \ldots + \frac{1}{x_n}\right)$$

$$\geq n \sqrt[n]{x_1 x_2 \ldots x_n} \cdot \frac{n}{\sqrt[n]{x_1 x_2 \ldots x_n}}$$

$$\geq n \sqrt[n]{x_1 x_2 \ldots x_n} \cdot \frac{n}{\sqrt[n]{x_1 x_2 \ldots x_n}}$$

$$= n^2$$

$$\geq 16$$

We obtained a contradiction.

Problem 6 (Combinatorics)

There is a pile of 2019 stones on a table. You are allowed to perform the following operation: you choose one of the piles containing more than 1 stone, throw away one stone from that pile and divide the pile into two smaller (not necessarily equal) piles. Is it possible to reach a situation in which all the piles on the table contain exactly 7 stones?

Solution

Answer: it is impossible.

Let us assume that it is possible to have exactly n piles with 7 stones each. Notice that after each operation the total number of stones in the piles decreases by 1, while the total number of piles increases by 1. Therefore, the sum of the total number of stones and the total number of piles is invariant.

In the beginning the sum is equal to $2019 + 1 = 2020$. In the end the sum is equal to

$$7n + n = 8n$$

Since 2020 is not divisible by 8, we obtained a contradiction.

CHAPTER 4

INTERMEDIATE LEVEL EXAM OF 2020

Problem 1 (Number Theory)

The number $x^2 + y^2$ is divisible by 3. Is it possible for y to be equal to 2020?

Solution

Answer: no.

Let us assume that $y = 2020$, which implies that $y \equiv 1 \pmod 3$. Since $x^2 + y^2$ is divisible by 3, then

$$x^2 + y^2 \equiv 0 \pmod 3$$
$$x^2 + (1)^2 \equiv 0 \pmod 3$$
$$x^2 \equiv -1 \pmod 3$$
$$x^2 \equiv 2 \pmod 3$$

However x^2 only gives remainders 0 or 1 when divided by 3.

We obtained a contradiction.

Junior, Intermediate and Senior Math Olympiads
by Roman Kvasov, Ph.D.

85

Problem 2 (Geometry)

Let ABC be a triangle with $AB > AC$ and let k be its circumcircle. The line tangent to the circle k at the point A intersects the line BC at the point P. Let m be a line passing through the point P and intersecting the sides AB and AC at the points D and E respectively, such that $AD = AE$. Show that the line m is the angle bisector of the angle $\angle APB$.

Solution

The solution presented below refers to Figure 4.1.

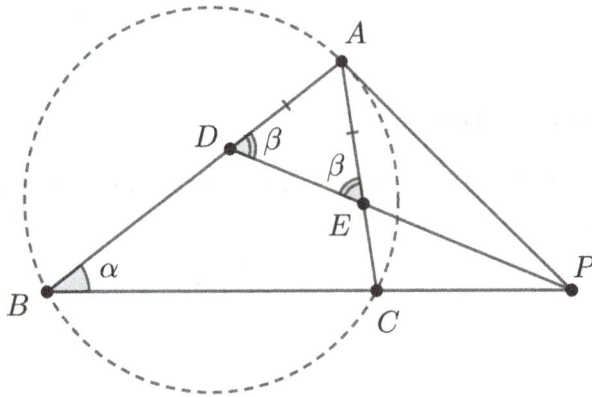

Figure 4.1   Line PE is the angle bisector of the angle $\angle APB$ in Problem 2.

Let $\angle ABC = \alpha$ and $\angle ADE = \beta$.

Since $AD = AE$, then the triangle ADE is isosceles and

$$\angle AED = \angle ADE = \beta$$

Since AP is tangent to k, then

$$\angle CAP = \angle CBA = \alpha$$

We have

$$\angle AEP = 180° - \beta$$

and thus

$$\angle EPA = \beta - \alpha$$

Also
$$\angle BDP = 180° - \beta$$

and thus
$$\angle BPD = \beta - \alpha$$

Therefore, we have that $\angle APD = \angle BPD$ and the line m is the angle bisector of the angle $\angle APB$.

Problem 3 (Algebra)

Positive real numbers a, b and c are such that $abc = 1$. Prove that

$$6a + 3b^2 + 2c^3 \geq 11$$

Solution

Let us rewrite the left-hand side and apply AM-GM to 11 terms:

$$a + a + a + a + a + a + b^2 + b^2 + b^2 + c^3 + c^3 \geq \sqrt[11]{a^6 b^6 c^6} = 11$$

Problem 4 (Number Theory)

Prove that there is an infinite number of triples (x, y, z) of positive integers, such that

$$x^2 + y^2 + z^2$$

is a perfect square, such that the greatest common divisor of the three numbers x, y, z is 1.

Solution

It is enough to put

$$x = 1$$
$$y = 2n$$
$$z = 2n^2$$

where n is any positive integer.

Indeed, the greatest common divisor of the three numbers x, y, z is 1 and we have

$$x^2 + y^2 + z^2 = (1)^2 + (2n)^2 + (2n^2)^2$$
$$= 1 + 4n^2 + 4n^4$$
$$= (1 + 2n^2)^2$$

as desired.

Problem 5 (Combinatorics)

In the beginning there is a box with 100 danish cookies. Lucas and Summer are playing the following game. Lucas goes first and eats no more than half of the cookies present in the box. Then goes Summer and eats no more than half of the cookies that are left in the box, etc. At each turn the players can only eat no more than half of the cookies left in the box until one of them cannot eat any cookies and loses the game. Who has a winning strategy?

Solution

Answer: Lucas has a winning strategy.

Working backwards we can find the "winning positions" for Lucas in this game to be 1, 4, 7, 15, 31 and 63.

If Lucas leaves Summer 1 cookie, then he wins. If Lucas leaves Summer 2 cookies, then he loses.

If Lucas leaves Summer 3 cookies, then he wins. It is not hard to see that if Lucas leaves Summer 4, 5 or 6 cookies, then he will loose, as Summer will be able to make a move and leave Lucas 3 cookies.

If Lucas leaves Summer with 7 cookies, then he wins. Similarly, if Lucas leaves Summer any of the 8, 9, ... , 14 cookies, then he will loose and if Lucas leaves Summer 15 cookies, then he will win.

Similarly, if Lucas leaves Summer 31 and 63 cookies, then he will win. Therefore, Lucas's first move should be to eat 37 cookies, leaving Summer 63 cookies and then 31, 15, 7, 3 and 1 cookie.

Problem 6 (Algebra)

Given that

$$\frac{a}{b+c} = \frac{b}{c+a} = \frac{c}{a+b}$$

Find all possible values of the expression

$$S = \frac{a+b}{c} + \frac{b+c}{a} + \frac{c+a}{b}$$

Solution

Answer: $S = -3$ or $S = 6$.

From the first equality we have

$$a(a+c) = b(b+c)$$
$$a^2 + ac = b^2 + bc$$
$$a^2 + ac - b^2 - bc = 0$$
$$\left(a^2 - b^2\right) + (ac - bc) = 0$$
$$(a-b)(a+b) + c(a-b) = 0$$
$$(a-b)(a+b+c) = 0$$

From the second equality we have

$$b(a+b) = c(a+c)$$
$$ab + b^2 = ac + c^2$$
$$ab + b^2 - ac - c^2 = 0$$
$$\left(b^2 - c^2\right) + (ab - ac) = 0$$
$$(b-c)(b+c) + a(b-c) = 0$$
$$(b-c)(a+b+c) = 0$$

Without loss of generality, we will proceed by doing the following casework:

- If $a+b+c = 0$, then $a+b = -c$, $b+c = -a$, $a+c = -b$ and

$$S = -1 - 1 - 1 = -3$$

- If $a+b+c \neq 0$, then $a-b = 0$ and $b-c = 0$, which implies that $a = b = c$ and

$$S = 2 + 2 + 2 = 6$$

CHAPTER 5

INTERMEDIATE LEVEL EXAM OF 2021

Problem 1 (Algebra)

Real numbers a, b and c satisfy the equalities

$$a - b = c$$
$$a^2 - b^2 = c^2$$

Prove that at least one of the numbers is zero.

Solution

Let us rewrite the second equation and substitute $a - b = c$:

$$a^2 - b^2 = c^2$$
$$(a - b)(a + b) = c^2$$
$$c(a + b) = c^2$$
$$c(a + b) - c^2 = 0$$
$$c(a + b - c) = 0$$

From here either $c = 0$ or $a + b - c = 0$.

Junior, Intermediate and Senior Math Olympiads
by Roman Kvasov, Ph.D.

91

We will proceed by doing the following casework:

- If $c = 0$, then one of the numbers is zero, as desired.

- If $a + b - c = 0$, then substituting $c = a - b$ from the first equation we have

$$a + b - c = 0$$
$$a + b - (a - b) = 0$$
$$a + b - a + b = 0$$
$$2b = 0$$
$$b = 0$$

and one of the numbers is zero, as desired

Problem 2 (Number Theory)

Let m and n be integer numbers, such that $2022m^2 + 2023n^2$ is divisible by 4. Is it true that $2023m^2 + 2022n^2$ is also divisible by 4?

Solution

Answer: yes.

Let us consider the following table of residues modulo 4:

$x \pmod 4$	$x^2 \pmod 4$
0	$(0)^2 \equiv 0$
1	$(1)^2 \equiv 1$
2	$(2)^2 \equiv 4 \equiv 0$
3	$(3)^2 \equiv 9 \equiv 1$

From here we see that a number that is a square of an integer is congruent to 0 or 1 modulo 4. This implies that the number $2022m^2$ is congruent to 0 or 2 modulo 4, while the number $2023n^2$ is congruent to 0 or 3 modulo 4. Therefore, the number $2022m^2 + 2023n^2$ is divisible by 4 if and only if both numbers m^2 and n^2 give remainder 0 modulo 4. Consequently, we have

$$2023m^2 + 2022n^2 = 2023(0) + 2022(0) \equiv 0 \pmod 4$$

as desired.

Problem 3 (Geometry)

Given a parallelogram $ABCD$ and the point K, such that $AK = BD$. Let M be the midpoint of the segment CK. Prove that $\angle BMD = 90°$.

Solution

The solution presented below refers to Figure 5.1.

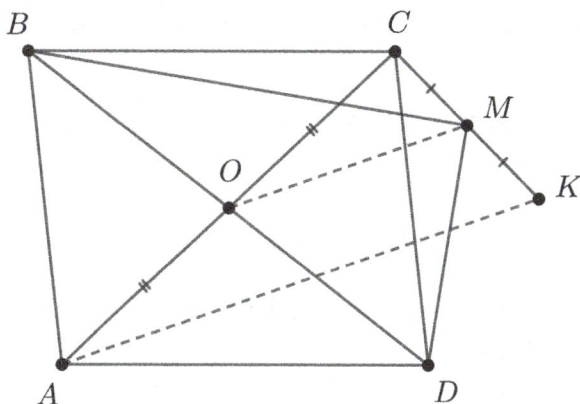

Figure 5.1 MO is a median of the triangle BMD in Problem 3.

Let O be the point of intersection of the diagonals AC and BD.

If the point K belongs to the line AC, then $OM = \frac{1}{2}AK$.

If the point K does not belong to the line AC, then OM is the midsegment of the triangle ACK and $OM = \frac{1}{2}AK$.

In the triangle BMD, the median MO is iqual to the half of the side BD and, therefore, we have $\angle BMD = 90°$.

Problem 4 (Combinatorics)

2021 points are chosen on the line containing the segment AB, all lying outside of segment AB. Prove that the sum of the distances from these points to the point A is not equal to the sum of the distances from these points to the point B.

Solution

Let x be the length of the segment AB and let X_1, X_2, ... , X_m be the points lying on the left from A and Y_1, Y_2, ... , Y_n be the points lying on the right from B. Notice that since $m + n = 2021$, then m and n are of different parity.

Let S_A and S_B be sums of the distances from all the points to the points A and B, respectively. Let us assume that the sum of the distances from these points to the points A and B are equal, i.e. $S_A = S_B$.

Let us define the quantity S as follows:

$$S = X_1 A + \ldots + X_m A + Y_1 B + \ldots + Y_n B$$

Then for S_A we have

$$
\begin{aligned}
S_A &= X_1 A + \ldots + X_m A + Y_1 A + \ldots + Y_n A \\
&= X_1 A + \ldots + X_m A + (Y_1 B + x) + \ldots + (Y_n B + x) \\
&= X_1 A + \ldots + X_m A + Y_1 B + \ldots + Y_n B + nx \\
&= S + nx
\end{aligned}
$$

Similarly for S_B we have

$$
\begin{aligned}
S_B &= X_1 B + \ldots + X_m B + Y_1 B + \ldots + Y_n B \\
&= (X_1 A + x) + \ldots + (X_m A + x) + Y_1 B + \ldots + Y_n B \\
&= X_1 A + \ldots + X_m A + mx + Y_1 B + \ldots + Y_n B \\
&= S + mx
\end{aligned}
$$

From here $S_A = S_B$ implies that

$$
\begin{aligned}
S_A &= S_B \\
S + nx &= S + mx \\
nx &= mx \\
n &= m
\end{aligned}
$$

which is impossible since m and n are of different parity. We obtained a contradiction.

Problem 5 (Number Theory)

Determine all positive integers $n \geq 2$ that can be written as

$$n = k^2 + d^2$$

where k is the smallest divisor of n greater than 1 and d is some other divisor of n.

Solution

Answer: $n = 20$.

Notice that since k is the smallest divisor of n greater than 1, then k is prime.

Since $d|d^2$ and $d|n$, then $d|k^2$. Since $d \neq k$, then $d = 1$ or $d = k^2$.

We will proceed by doing the following casework:

- Let us assume that $d = 1$. Then we have

$$n = k^2 + 1$$

  and since $k|n$ and $k|k^2$, then $k|1$. Contradiction.

- Let us assume that $d = k^2$. We have

$$n = k^2 + k^4$$

  and since $k^2 + k^4$ is even, then $k = 2$ and $n = 20$.

Problem 6 (Algebra)

Given the positive real numbers a, b, c. Prove the inequality

$$\frac{a^2b}{a+c} + \frac{b^2c}{b+a} + \frac{c^2a}{c+b} \geq \frac{ab+bc+ac}{2}$$

Solution

We will use the cyclic sums notation.

The inequality that is needed to be proven can be written as

$$\sum_{cyc} \frac{a^2 b}{a+c} \geq \frac{ab + bc + ac}{2}$$

Let us rewrite the left-hand side and apply the Titu's Lemma:

$$\sum_{cyc} \frac{a^2 b}{a+c} = \sum_{cyc} \frac{a^2 b^2}{ab + bc} \geq \frac{(ab + bc + ac)^2}{2(ab + bc + ac)} = \frac{ab + bc + ac}{2}$$

as desired.

CHAPTER 6

INTERMEDIATE LEVEL EXAM OF 2022

Problem 1

Every second the numbers a, b, c are substituted for the numbers $a + b - c$, $a + c - b$, $b + c - a$. Is it possible to obtain 2021, 2022, 2023 from 2019, 2021, 2024 in some order?

Solution

Answer: it is impossible.

Let us assume that it is possible to obtain the numbers 2021, 2022, 2023 from the numbers 2019, 2021, 2024. At a certain moment in time, let the numbers on the board be a, b and c. Let us consider the quantity Q defined as

$$Q(a, b, c) = a + b + c$$

and track how it changes under the described operation. After the operation is performed the quantity Q becomes

$$Q(a + b - c, a + c - b, b + c - a) = (a + b - c) + (a + c - b) + (b + c - a)$$
$$= a + b + c$$

This implies that the quantity Q is invariant under the described operation.

Junior, Intermediate and Senior Math Olympiads
by Roman Kvasov, Ph.D.

97

However, since in the beginning

$$Q(2019, 2021, 2024) = 2019 + 2021 + 2024 = 6064$$

and in the end

$$Q(2021, 2022, 2023) = 2021 + 2022 + 2023 = 6066$$

then we obtained a contradiction.

Problem 2 (Number Theory)

Find all distinct prime numbers p and q, such that the number

$$(p + q)^2 + (p - q)^2$$

is a multiple of pq.

Solution

Answer: there are no such prime numbers.

Let us assume that such prime numbers exist. Start by noticing that

$$(p + q)^2 + (p - q)^2 = 2p^2 + 2q^2$$

Since p divides $2p^2 + 2q^2$, then p will also divide $2q^2$. However, since $p \neq q$, then we conclude that $p = 2$.

Similarly, we have that since q divides $2p^2 + 2q^2$, then q will also divide $2p^2$. However, since $p \neq q$, then we conclude that $q = 2$.

We obtained a contradiction with the condition that p and q are distinct prime numbers.

Problem 3 (Geometry)

Points P and Q are chosen respectively on the sides BC and CD of the square $ABCD$, such that $\angle PAQ = 45°$. The diagonal BD intersects the segments AP and AQ at the points M and N, respectively. Let O be the point of intersection of the lines PN and MQ. Prove that the line AO is perpendicular to the line PQ.

Solution

The solution presented below refers to Figure 6.1.

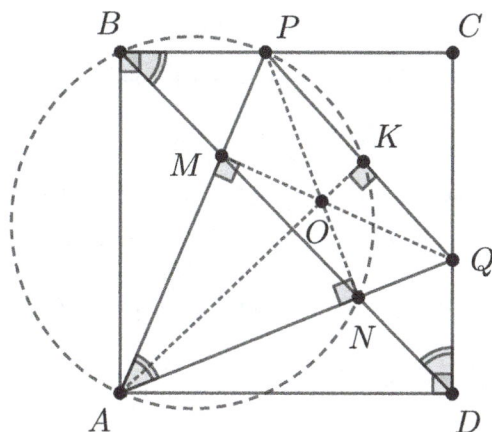

Figure 6.1   Quadrilaterals $ABPN$ and $ABQM$ are cyclic in Problem 3.

Start by noticing that since

$$\angle NBP = \angle NAP = 45°$$

then $ABPN$ is cyclic.

From here $\angle ANP = 90°$ and PN is the altitude in the triangle APQ.

Also, since

$$\angle MAQ = \angle MDQ = 45°$$

then $ABQM$ is cyclic.

From here $\angle AMQ = 90°$ and QM is the altitude in the triangle APQ.

This implies that O is the orthocenter of the triangle APQ and $AO \perp PQ$.

Problem 4 (Algebra)

Real numbers a, b, c and d satisfy the equation

$$a^2 + b^2 + c^2 + d^2 = a(b + c + d)$$

Prove that $a + b + c + d = 0$.

Solution

First, let us rewrite the equation as

$$a^2 + b^2 + c^2 + d^2 - ab - ac - ad = 0$$

Let us now multiply this equation by 4 and complete the squares:

$$4a^2 + 4b^2 + 4c^2 + 4d^2 - 4ab - 4ac - 4ad = 0$$
$$a^2 + \left(a^2 - 4ab + 4b^2\right) + \left(a^2 - 4ac + 4c^2\right) + \left(a^2 - 4ad + 4d^2\right) = 0$$
$$a^2 + (a - 2b)^2 + (a - 2c)^2 + (a - 2d)^2 = 0$$

This immediately implies that $a = 0$, and, consequently, $b = 0$, $c = 0$ and $d = 0$.

From here $a + b + c + d = 0$ as desired.

Problem 5 (Combinatorics)

9999 squares of size 2×2 are cut off from the board 299×299. Is it always possible to cut off one more 2×2 square?

Solution

Answer: yes, it is possible.

Let us paint 10000 squares of size 2×2 in black color in such way so the black squares are 1 square apart form each other and they do not share a common point (see Figure 6.2).

Notice that when we cut a 2×2 square it can only affect at most one black square. Therefore at most 9999 of these squares will be affected. By Pigeonhole Principle there exists a black square that will not be affected and we will be able to cut it off.

Problem 6 (Algebra)

Given positive real numbers a, b and c, such that

$$ab + bc + ca = abc$$

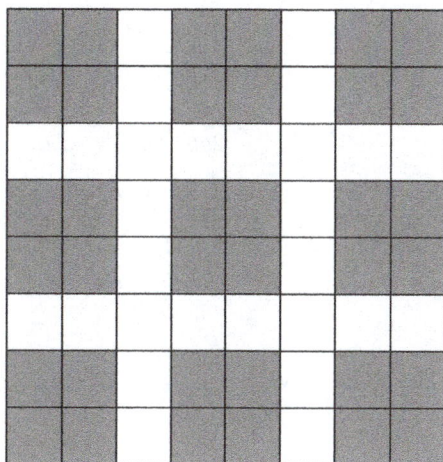

Figure 6.2   10000 squares of size 2×2 are painted in black in Problem 5.

Prove that

$$\frac{a+b}{a^2+b^2} + \frac{b+c}{b^2+c^2} + \frac{c+a}{c^2+a^2} \leq 1$$

Solution

Let us start by rewriting the given equality by dividing both sides by abc:

$$ab + bc + ca = abc$$

$$\frac{ab + bc + ca}{abc} = \frac{abc}{abc}$$

$$\frac{ab}{abc} + \frac{bc}{abc} + \frac{ca}{abc} = 1$$

$$\frac{1}{c} + \frac{1}{a} + \frac{1}{b} = 1$$

Let us apply the AM-GM Inequality to each of the following denominators:

$$a^2 + b^2 \geq 2\sqrt{a^2b^2} = 2ab$$

$$b^2 + c^2 \geq 2\sqrt{b^2c^2} = 2bc$$

$$c^2 + a^2 \geq 2\sqrt{c^2a^2} = 2ca$$

From here

$$\frac{a+b}{a^2+b^2} + \frac{b+c}{b^2+c^2} + \frac{c+a}{c^2+a^2} \leq \frac{a+b}{2ab} + \frac{b+c}{2bc} + \frac{c+a}{2ca}$$

$$= \frac{a}{2ab} + \frac{b}{2ab} + \frac{b}{2bc} + \frac{c}{2bc} + \frac{c}{2ca} + \frac{a}{2ca}$$

$$= \frac{1}{2b} + \frac{1}{2a} + \frac{1}{2c} + \frac{1}{2b} + \frac{1}{2a} + \frac{1}{2c}$$

$$= \frac{1}{a} + \frac{1}{b} + \frac{1}{c}$$

$$= 1$$

which is what needed to be proven.

CHAPTER 7

INTERMEDIATE LEVEL EXAM OF 2023

Problem 1 (Combinatorics)

In a collection of 2023 coins, 202 of them are counterfeit. A counterfeit coin differs from the genuine coin in weight by exactly 1 gram. Peter possesses a balance scale that can only indicate the weight difference between objects placed in each pan. Peter selects one coin and seeks to determine, in a single weighing, whether it is counterfeit. Can he accomplish this?

Solution

Answer: yes, he can accomplish this.

Start by noticing that there are $2023 - 202 = 1821$ genuine coins, which is an odd number.

Let us assume that the selected coin is genuine. Then the rest 2022 coins contain exactly 1820 genuine coins and 202 counterfeit coins. Let us divide these 2022 coins randomly into two sets of 1011 coins each and weigh these sets.

In this case the balance scale will indicate their weight difference, which will be an even number. Indeed, since the number of counterfeit coins is even, then the difference of their contributions to each set will be even.

Junior, Intermediate and Senior Math Olympiads
by Roman Kvasov, Ph.D.

103

Let us assume that the selected coin is counterfeit. Then the rest 2022 coins contain exactly 1821 genuine coins and 201 counterfeit coins. Let us divide these 2022 coins randomly into two sets of 1011 coins each and weigh these sets.

In this case the balance scale will indicate their weight difference, which will be an odd number. Indeed, since the number of counterfeit coins is odd, then the difference of their contributions to each set will be odd.

Therefore, Peter can determine, in a single weighing, whether a chosen coin is counterfeit by following the strategy:

- Divide the remaining 2022 coins randomly into two sets of 1011 coins each.

- Weigh the sets.

- If the weight difference indicate by the balance scale is even, the the chosen coin is genuine, and counterfeit otherwise.

Problem 2 (Geometry)

Let k_1 be a circle and l a line that intersects k_1 in two distinct points A and B. Let k_2 be another circle outside of k_1 that touches k_1 at C and l at D. Let T be the second intersection of k_1 and the line CD. Prove that $AT = TB$.

Solution

The solution presented below refers to Figure 7.1.

Let E be the point of intersection of the line l and the common tangents of the circles k_1 and k_2.

Let $\angle DCE = \alpha$ and $\angle BCE = \beta$. Then

$$\angle BDC = \alpha$$
$$\angle TAC = \alpha$$
$$\angle BAC = \beta$$

Form the triangle ACD we have that

$$\angle ACB = 180° - 2\alpha - 2\beta$$

Since $ABCT$ is cyclic then

$$\angle ATB = \angle ACB = 180° - 2\alpha - 2\beta$$

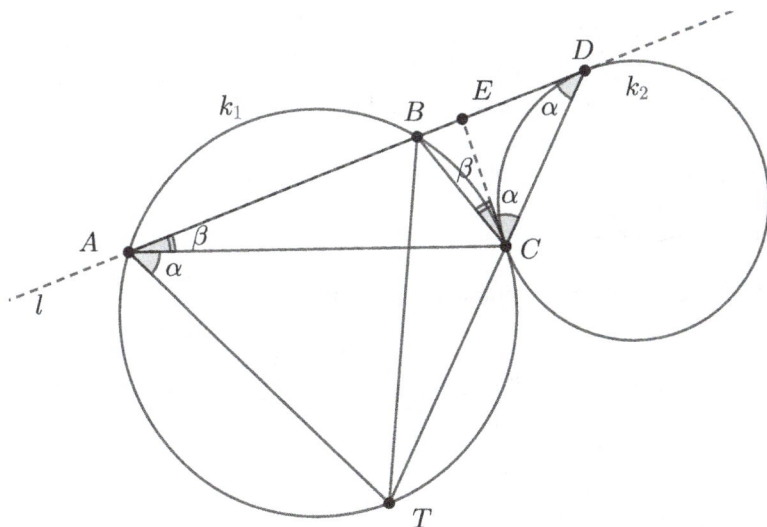

Figure 7.1   Triangle ABT is isosceles in Problem 2.

From the triangle ABT we have that

$$\angle ABT = 2\alpha + 2\beta$$

Therefore, the triangle ABT is isosceles and $TA = TB$, as desired.

Problem 3 (Number Theory)

Positive integers a, b and d are such that the number $a + b$ is divisible by d, and the number ab is divisible by d^2. Prove that if d is odd, then each of the numbers a and b is divisible by d.

SOLUTION

Since d divides $a + b$, then d^2 divides $(a + b)^2$. However, since d^2 divides ab and

$$(a + b)^2 = a^2 + 2ab + b^2$$

then d^2 also divides $a^2 + b^2$. Therefore d^2 will also divide the expression

$$a^2 - 2ab + b^2 = (a - b)^2$$

which means that d divides $(a - b)$. Furthermore, we have that d should divide

$$(a + b) + (a - b) = 2a$$

and d should also divide

$$(a + b) - (a - b) = 2b$$

Since d is odd, then d divides a and b, as desired.

Problem 4 (Algebra)

Let a, b, c, d be positive real numbers. Show that

$$\frac{4a^2}{b + 3c} + \frac{4b^2}{c + 3d} + \frac{4c^2}{d + 3a} + \frac{4d^2}{a + 3b} \geq a + b + c + d$$

Solution

Let us apply the Titu's Lemma for the numbers $(2a, 2b, 2c, 2d)$ and the numbers $(b + 3c, c + 3d, d + 3a, a + 3b)$:

$$\frac{4a^2}{b + 3c} + \frac{4b^2}{c + 3d} + \frac{4c^2}{d + 3a} + \frac{4d^2}{a + 3b} \geq \frac{4(a + b + c + d)^2}{4(a + b + c + d)} = a + b + c + d$$

as desired.

Problem 5 (Number Theory)

Let p be a prime number and a, b, c and n positive integers with $a, b, c < p$, such that p^2 divides each of the numbers $a + (n - 1)b$, $b + (n - 1)c$ and $c + (n - 1)a$. Show that n is not a prime number.

Solution

Assume that n is prime.

We will solve the problem by doing the following casework:

- If $p = 2$, then $a = b = c = 1$. From here $4|n$ and n is not a prime number.
- If $p \geq 3$, then p^2 should divide the sum of $a + (n - 1)b$, $b + (n - 1)c$ and $c + (n - 1)a$:

$$(a + (n - 1)b) + (b + (n - 1)c) + (c + (n - 1)a) = n(a + b + c)$$

However, we have

$$a + b + c < p + p + p = 3p \leq p^2$$

Therefore $p|n$ and since n is prime then $n = p$. On another hand

$$a + (n-1)b < p + (p-1)p = p^2$$

and, therefore, cannot be divisible by p^2. We obtained a contradiction.

Problem 6 (Algebra)

Find all functions $f : \mathbb{R} \to \mathbb{R}$, such that for all $x, y \in \mathbb{R}$

$$f\left(x + y^2\right) = f\left(x^2\right) + f\left(2023y\right)$$

Solution

Answer: $f(x) = 0$ for all $x \in \mathbb{R}$.

Let us substitute y for $-y$ into the original functional equation:

$$f\left(x + y^2\right) = f\left(x^2\right) + f\left(-2023y\right)$$

Combining it with the original functional equation we obtain that

$$f\left(-2023y\right) = f\left(2023y\right)$$

Substituting $2023y$ for a new variable t we have

$$f(-t) = f(t)$$

Since t can be any real number, then the function f is odd on $\mathbb{R}$.

Let us substitute x for $-x$ into the original functional equation:

$$f\left(-x + y^2\right) = f\left(x^2\right) + f\left(2023y\right)$$

Combining it with the original functional equation we obtain that

$$f\left(x + y^2\right) = f\left(-x + y^2\right)$$

Substituting x for y^2 into the last equation we have

$$f\left(2y^2\right) = 0$$

Substituting $2y^2$ for a new variable t we have

$$f(t) = 0$$

Since t can be any nonnegative real number, then the function f is a constant zero function on $[0, +\infty)$. Since f is odd on $\mathbb{R}$, then

$$f(-t) = f(t) = 0$$

From here $f(x) = 0$ for all $x \in \mathbb{R}$.

CHAPTER 8

INTERMEDIATE LEVEL EXAM OF 2024

Problem 1 (Geometry)

Two circles intersect at the points P and Q and the line l intersects the circles at the points A, B, C, D (in that order). Show that $\angle APB = \angle CQD$.

Solution

The solution presented below refers to Figure 8.1.

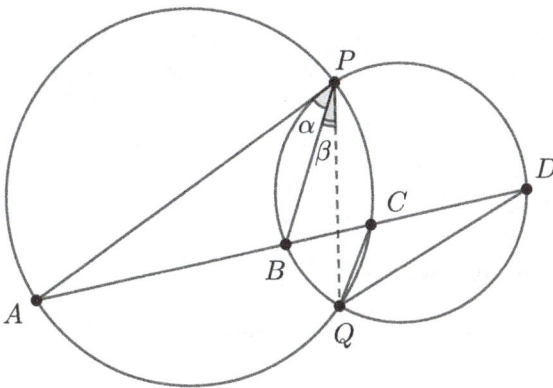

Figure 8.1   Quadrilaterals $APCQ$ and $BPDQ$ are cyclic in Problem 1.

Junior, Intermediate and Senior Math Olympiads
by Roman Kvasov, Ph.D.

109

Let us put $\angle APB = \alpha$ and $\angle BPQ = \beta$.

Since $APCQ$ is cyclic, then

$$\angle ACQ = \angle APQ = \angle APB + \angle BPQ = \alpha + \beta$$

Since $BPDQ$ is cyclic, then

$$\angle BDQ = \angle BPQ = \beta$$

Notice that the angle $\angle ACQ$ is the external angle of the triangle QCD and therefore

$$\angle CQD + \angle CDQ = \angle ACQ$$

From here

$$\angle CQD = \angle ACQ - \angle CDQ = \angle ACQ - \angle BDQ = (\alpha + \beta) - \beta = \alpha$$

Therefore $\angle APB = \angle CQD$, as desired.

Problem 2 (Number Theory)

Given some natural number n. Show that the numbers $n(n-1)$ and $(n+1)^2$ have different sums of their digits.

Solution

Let $S(k)$ represent the sum of digits of the number k. Let us assume that the numbers $n(n-1)$ and $(n+1)^2$ have equal sums of their digits, i.e.

$$S\left((n+1)^2\right) = S\left(n(n-1)\right)$$

or equivalently

$$S\left((n+1)^2\right) - S\left(n(n-1)\right) = 0$$

Notice that for every positive integer k:

$$k \equiv S(k) \pmod 3$$

and we have

$$S\left((n+1)^2\right) - S\left(n(n-1)\right) \equiv (n+1)^2 - n(n-1) = 3n+1 \equiv 1 \pmod 3$$

which leads to a contradiction.

Problem 3 (Algebra)

Positive numbers a, b and c satisfy the equalities

$$a + b = c$$
$$a^3 + a^2b + a^2 = c + 1$$

Prove that at least one of the numbers is equal to 1.

Solution

It will be enough to prove that $a = 1$.

Let us factor out the common factor a^2 in the second equation and substitute $a + b$ for c from the first equation:

$$a^3 + a^2b + a^2 = c + 1$$
$$a^2(a + b + 1) = c + 1$$
$$a^2(c + 1) = c + 1$$
$$a^2(c + 1) - (c + 1) = 0$$
$$(c + 1)\left(a^2 - 1\right) = 0$$
$$(c + 1)(a - 1)(a + 1) = 0$$

Notice that since a and c are positive numbers, then the last equality implies that $a = 1$, as desired.

Problem 4 (Combinatorics)

There are 2017 red, 2019 green and 2021 blue chameleons on the Rainbow Island. When two chameleons of different colors meet they both change their color to the third one (for instance, if blue and green meet they will both become red). Is it possible that after some time all the chameleons are of the same color?

Solution

Answer: it is impossible.

Let (r, g, b) be the triple of remainders modulo 3 of the numbers or red, green and blue chameleons, respectively. Notice that if two chameleons meet, then the triple of remainders becomes $(r-1, g-1, b-1)$. Initially the triple of remainders is $(1, 0, 2)$ and it changes as follows:

$$(1, 0, 2) \to (0, 2, 1) \to (2, 1, 0) \to (1, 0, 2) \to \ldots$$

If after some time all the chameleons are of the same color, then the triple would be $(0, 0, 0)$, $(1, 1, 1)$ or $(2, 2, 2)$, which is impossible.

Problem 5 (Number Theory)

Find all pairs (a, b) of integer numbers that satisfy the condition

$$a^2 + b^2 = a + b + ab$$

Solution

Answer: $(a, b) \in \{(2, 2), (0, 0), (1, 2), (1, 0), (2, 1), (0, 1)\}$.

Let us rewrite the equality and complete the squares as follows:

$$a^2 + b^2 = a + b + ab$$
$$a^2 + b^2 - a - b - ab = 0$$
$$2a^2 + 2b^2 - 2a - 2b - 2ab = 0$$
$$\left(a^2 - 2ab + b^2\right) + \left(a^2 - 2a + 1\right) + \left(b^2 - 2b + 1\right) = 2$$
$$(a - b)^2 + (a - 1)^2 + (b - 1)^2 = 2$$

From here exactly two of the squares are equal to 1 and one of the squares is equal to 0. We will proceed by doing the following casework:

- If $a - b = 0$, then $a = b = 0$ or $a = b = 2$.

- If $a - 1 = 0$, then $a = 1$ and $b = 0$ or $b = 2$.

- If $b - 1 = 0$, then $b = 1$ and $a = 0$ or $a = 2$.

Therefore we have

$$(a, b) \in \{(2, 2), (0, 0), (1, 2), (1, 0), (2, 1), (0, 1)\}$$

Problem 6 (Algebra)

Prove that for $n \in \mathbb{N}$, the inequality holds:

$$\frac{1}{2 \cdot n} + \frac{2}{3 \cdot n} + \ldots + \frac{n-1}{n \cdot n} \geq \frac{1}{\sqrt[n]{n+1}} - \frac{1}{n+1}$$

Solution

Let us multiply both sides of the given inequality by n and rewrite it as follows:

$$\frac{1}{2 \cdot n} + \frac{2}{3 \cdot n} + \ldots + \frac{n-1}{n \cdot n} \geq \frac{1}{\sqrt[n]{n+1}} - \frac{1}{n+1}$$

$$\frac{1}{2} + \frac{2}{3} + \ldots + \frac{n-1}{n} \geq \frac{n}{\sqrt[n]{n+1}} - \frac{n}{n+1}$$

$$\frac{1}{2} + \frac{2}{3} + \ldots + \frac{n-1}{n} + \frac{n}{n+1} \geq \frac{n}{\sqrt[n]{n+1}}$$

Let us now apply the AM-GM Inequality to the left-hand side:

$$\frac{1}{2} + \frac{2}{3} + \cdots + \frac{n-1}{n} + \frac{n}{n+1} \geq n \sqrt[n]{\frac{1}{2} \cdot \frac{2}{3} \cdot \ldots \cdot \frac{n-1}{n} \cdot \frac{n}{n+1}}$$

$$= n \sqrt[n]{\frac{1}{\cancel{2}} \cdot \frac{\cancel{2}}{\cancel{3}} \cdot \ldots \cdot \frac{\cancel{n-1}}{\cancel{n}} \cdot \frac{\cancel{n}}{n+1}}$$

$$= \frac{n}{\sqrt[n]{n+1}}$$

as desired.

CHAPTER 9

TOPICS FOR INTERMEDIATE MATH OLYMPIADS

Factorization of n-th Powers

- For any positive integer power n:

$$a^n - b^n = (a - b) \sum_{k=0}^{n-1} a^{n-1-k} b^k$$

$$= (a - b) \left(a^{n-1} + a^{n-2}b + \ldots + ab^{n-2} + b^{n-1} \right)$$

- For positive integer odd powers n:

$$a^n + b^n = (a + b) \sum_{k=0}^{n-1} (-1)^k a^{n-1-k} b^k$$

$$= (a + b) \left(a^{n-1} - a^{n-2}b + \ldots - ab^{n-2} + b^{n-1} \right)$$

Binomial Formula

For any positive integer power n and real a and b:

$$(a + b)^n = \sum_{k=0}^{n} \binom{n}{k} a^{n-k} b^k$$

where the binomial coefficients are given as

$$\binom{n}{k} = \frac{n!}{k!(n-k)!}$$

Junior, Intermediate and Senior Math Olympiads
by Roman Kvasov, Ph.D.

115

Vieta's Formulas for Quadratic Polynomial

If x_1 and x_2 are the roots of the polynomial

$$f(x) = ax^2 + bx + c$$

where $a \neq 0$, then the following equalities hold

$$x_1 + x_2 = -\frac{b}{a}$$

$$x_1 \cdot x_2 = \frac{c}{a}$$

Vieta's Formulas for Cubic Polynomial

If x_1, x_2 and x_3 are the roots of the polynomial

$$f(x) = x^3 + ax^2 + bx + c$$

then the following equalities hold

$$x_1 + x_2 + x_3 = -a$$

$$x_1 \cdot x_2 + x_2 \cdot x_3 + x_3 \cdot x_1 = b$$

$$x_1 \cdot x_2 \cdot x_3 = -c$$

Inequality of Arithmetic and Geometric Means

For the positive real numbers $(a_1, a_2, ..., a_n)$ the following inequality holds:

$$\frac{a_1 + a_2 + ... + a_n}{n} \geq \sqrt[n]{a_1 \cdot a_2 \cdot ... \cdot a_n}$$

The equality holds when all the variables a_i are equal.

Titu's Lemma

For positive real numbers $(a_1, a_2, ..., a_n)$ and $(b_1, b_2, ..., b_n)$ the following inequality holds:

$$\frac{a_1^2}{b_1} + \frac{a_2^2}{b_2} + ... + \frac{a_n^2}{b_n} \geq \frac{(a_1 + a_2 + ... + a_n)^2}{b_1 + b_2 + ... + b_n}$$

The equality holds when the variables are proportional, i.e. if there exists a constant t such that $a_i = tb_i$ for all $i = 1, 2, ..., n$.

Cauchy-Schwarz Inequality

For the real numbers $(a_1, a_2, \ldots, a_n)$ and $(b_1, b_2, \ldots, b_n)$ the following inequality holds:

$$(a_1 b_1 + a_2 b_2 + \ldots + a_n b_n)^2 \leq \left(a_1^2 + a_2^2 + \ldots + a_n^2\right) \cdot \left(b_1^2 + b_2^2 + \ldots + b_n^2\right)$$

The equality holds when the variables are proportional, i.e. if there exists a constant t such that $a_i = tb_i$ for all $i = 1, 2, \ldots, n$.

Greatest Common Divisor

Given the positive integer numbers a and b. Number d is called the **greatest common divisor** of the numbers a and b, if it is the largest positive integer that divides both a and b. **Greatest common divisor** of the numbers a and b is usually written as $\gcd(a, b)$.

Properties of Greatest Common Divisor

1. **Identity Property**. For positive integer a

$$\gcd(a, 1) = 1$$

2. **Reflexive Property**. For positive integer a

$$\gcd(a, a) = a$$

3. **Commutative Property**. For positive integers a and b

$$\gcd(a, b) = \gcd(b, a)$$

4. **Associative Property**. For positive integers a, b and c

$$\gcd\left(a, \gcd(b, c)\right) = \gcd\left(\gcd(a, b), c\right)$$

5. **Multiplicative Property**. For positive integer b, and relatively prime numbers a_1 and a_2

$$\gcd\left(a_1 \cdot a_2, b\right) = \gcd\left(a_1, b\right) \cdot \gcd\left(a_2, b\right)$$

6. **Common Factor Property**. For positive integers a, b and m

$$\gcd(m \cdot a, m \cdot b) = m \cdot \gcd(a, b)$$

7. **Remainder Property**. If r is the remainder of the division of the positive integer a by the the positive integer b, then

$$\gcd(a, b) = \gcd(b, r)$$

Euclidean Algorithm

Euclidean Algorithm is a method for finding the greatest common divisor of two integers. It relies on the following list of steps:

1.   If $a > b$, then divide a by b and let r be the remainder.

2.   Replace a with b and b with r.

3.   Repeat the division until the remainder becomes 0.

4.   The greatest common divisor is the last non-zero remainder.

Least Common Multiple

Given the positive integer numbers a and b. Number m is called the **least common multiple** of the numbers a and b, if it is the smallest positive integer that is divisible by both a and b. **Least common multiple** of the numbers a and b is usually written as $\mathrm{lcm}(a, b)$.

Properties of Least Common Multiple

1. **Identity Property**. For positive integer a

$$\mathrm{lcm}(a, 1) = a$$

2. **Reflexive Property**. For positive integer a

$$\mathrm{lcm}(a, a) = a$$

3. **Commutative Property**. For positive integers a and b

$$\mathrm{lcm}(a, b) = \mathrm{lcm}(b, a)$$

4. **Associative Property**. For positive integers a, b and c

$$\mathrm{lcm}\left(a, \mathrm{lcm}(b, c)\right) = \mathrm{lcm}\left(\mathrm{lcm}(a, b), c\right)$$

5. **Common Factor Property**. For positive integers a, b and m

$$\mathrm{lcm}(m \cdot a, m \cdot b) = m \cdot \mathrm{lcm}(a, b)$$

6. **Product Property**. For positive integers a, b and m

$$\gcd(a, b) \cdot \mathrm{lcm}(a, b) = a \cdot b$$

Chinese Remainder Theorem

If $m_1, m_2, \ldots, m_k$ are pairwise coprime positive integers, and $a_1, a_2, \ldots, a_k$ are any integers, then the system of congruences

$$x \equiv a_1 \pmod{m_1}$$
$$x \equiv a_2 \pmod{m_2}$$
$$\vdots \qquad \vdots$$
$$x \equiv a_k \pmod{m_k}$$

has a unique solution modulo $M = m_1 \cdot m_2 \cdot \ldots \cdot m_k$. In other words, all the solutions y of this system of congruences are given as

$$y \equiv x \pmod{M}$$

Bézout's Identity

Let a and b are positive integers and d is their greatest common divisor. Then there exist integer numbers x and y, such that

$$d = ax + by$$

Strategies for Games

- *No Strategy* – in these types of games one player always wins no matter what they do and no particular player have a strategy.

- *Symmetric Strategy* – in these types of games one player responds symmetrically to the other player's moves.

- *Winning Positions* – in these types of games one player after each move should try to land on a "winning position", i.e. the position that will make the other player lose.

Mathematical Induction

Mathematical Induction is used to prove statements that hold for all natural numbers n. It typically involves proving a base case (usually the statement is true for the smallest value, for example $n = 1$) and then showing that if the statement holds for some value k, it also holds for the next value $k + 1$. This allows us to conclude that the statement is true for all natural numbers.

Invariant

Invariant is a quantity that does not change under the described operation. Usually such quantities can be sums, products, sums of squares, or remainders by a modulus of the numbers under the consideration or some their subset.

Coloring Techniques

Coloring Techniques involve assigning two or more distinct colors to the objects in a problem. This approach often helps to isolate specific objects and discover useful properties of objects that share the same color.

Inscribed and Central Angles

Let the points A, B and C be the points on the circle with center O located such that B and O are on one side with respect to the line AC (see Figure 9.1). The $\angle ABC$ is called the **inscribed angle**, the $\angle AOC$ is called the **central angle** and

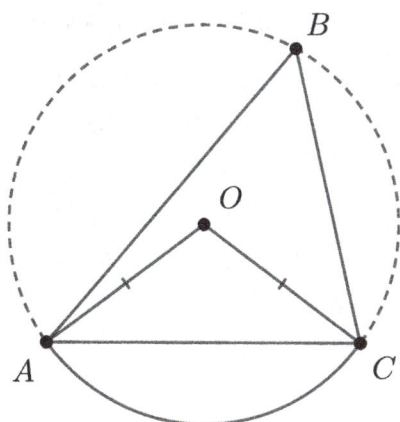

Figure 9.1   Angle $\angle ABC$ is inscribed and angle $\angle AOC$ is central.

the following equality holds:

$$\angle AOC = 2\angle ABC$$

Cyclic Quadrilaterals

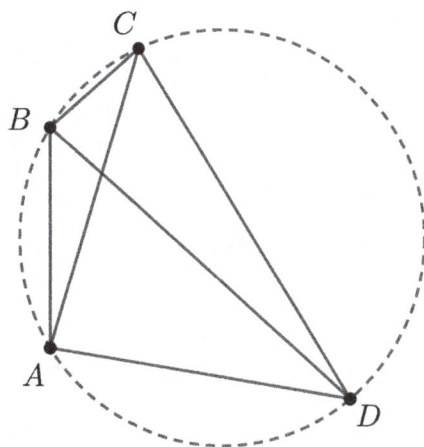

Figure 9.2   Points A, B, C and D lie on the same circle.

The points A, B, C and D lie on the same circle (or form a cyclic quadrilateral) if and only if at least one of following equalities holds (see Figure 9.2):

$$\angle ABD = \angle ACD$$
$$\angle BAC = \angle BDC$$
$$\angle BCA = \angle BDA$$
$$\angle CAD = \angle CBD$$
$$\angle ABC + \angle ADC = 180°$$
$$\angle BAD + \angle BCD = 180°$$

PART III

SENIOR MATH OLYMPIADS

Senior Level math olympiads are designed for students up to the 12th grade, providing a valuable opportunity for seasoned participants with extensive exposure to olympiad mathematics. These competitions encompass a broad array of advanced topics in algebra, geometry, combinatorics, and number theory, with the aim of challenging and deepening the understanding of mathematical concepts among participants.

In the realm of algebra, participants can expect to tackle multi-step problems involving the Weighted Inequality of Arithmetic and Geometric Means, Titu's Lemma, Bernoulli's, Chebyshev and Rearrangement inequalities, and their combinations with other algebraic techniques, along with proficiency in handling more difficult functional equations and some advanced methods, such as the use of vectors, complex numbers, and trigonometric substitutions.

Geometry problems at the Senior level often revolve around concepts such as similar triangles, power of a point, cyclic quadrilaterals, transformations, Ceva and Menelaus theorems, and more advanced results.

Combinatorics exposes participants to advanced concepts, including induction, invariants, monovariants, bijections, discrete continuity, extremal principle, recursion, and intricate coloring techniques. These problems not only evaluate mathematical reasoning but also cultivate profound creative thinking and involve complex multistep logical deduction.

In the field of number theory, Senior-level participants encounter more advanced topics such as Fermat's little theorem, Euler's totient function, Euler's theorem, Wilson's theorem, Dirichlet's theorem, quadratic residues, Legendre symbol, multiplicative order, multiplicative inverses, and employing proof by infinite descent.

To prepare effectively for Senior Math Olympiads, aspiring participants should focus on mastering these fundamental concepts. This can be achieved through a combination of regular practice, exposure to a variety of problems, and participation in mock exams. Developing a profound understanding of the basic principles in algebra, geometry, combinatorics, and number theory will empower students to approach olympiad problems with confidence and creativity. Additionally, seeking guidance from experienced mentors or teachers and collaborating with peers in problem-solving sessions can further enhance one's preparation for the challenges presented by Senior math olympiads.

CHAPTER 1

SENIOR LEVEL EXAM OF 2017

Problem 1 (Algebra)

Find all functions $f : \mathbb{R} \to \mathbb{R}$, such that for all $x, y \in \mathbb{R}$:

$$f(x + 2017y) = f(y^2) + 2017f(x)$$

Solution

Answer: $f(x) = 0$.

It is not hard to check that $f(x) = 0$ indeed satisfies the given functional equation.

Let us substitute $x = 0$ and $y = 0$ into the given functional equation:

$$f((0) + 2017(0)) = f((0)^2) + 2017f(0)$$
$$f(0) = f(0) + 2017f(0)$$
$$0 = 2017f(0)$$
$$0 = f(0)$$

Junior, Intermediate and Senior Math Olympiads
by Roman Kvasov, Ph.D.

125

Let us now substitute $y = 0$ into the given functional equation:

$$f(x + 2017(0)) = f((0)^2) + 2017f(x)$$
$$f(x) = f(0) + 2017f(x)$$
$$f(x) = 2017f(x)$$
$$0 = 2016f(x)$$
$$0 = f(x)$$

as desired.

Problem 2 (Number Theory)

Find the largest natural number S, such that all its digits are distinct and it becomes five times smaller if we erase its first digit.

Solution

Answer: 3750.

Let $S = \overline{aX}$, where the number X has n digits. Then $\overline{aX} = 5\overline{X}$ or equivalently

$$a \cdot 10^n + X = 5X$$

We will proceed by doing the following casework:

- If $n \geq 4$, then $n = k + 4$ for some nonnegative $k \in \mathbb{Z}$. We have

$$a \cdot 10^n + X = 5X$$
$$a \cdot 10^n = 4X$$
$$a \cdot 10^{k+4} = 4X$$
$$a \cdot 10^k \cdot 10^4 = 4X$$
$$a \cdot 10^k \cdot 2500 = X$$

  Therefore X ends with two zeros, which contradicts the fact that all digits of X are distinct.

- If $n \leq 3$, and X has 3 digits then $n = 3$ and we have

$$a \cdot 10^3 + X = 5X$$
$$a \cdot 10^3 = 4X$$
$$a \cdot 250 = X$$

Notice that for $a \geq 4$ the number X will have more than 3 digits and thus the maximum S is reached when $a = 3$. Indeed for $a = 3$ we have that $A = 750$ and $S = 3750$.

Problem 3 (Geometry)

Points P and Q are chosen inside an equilateral triangle ABC, such that P is located inside of the triangle AQB, $PQ = QC$ and $\angle PAQ = \angle PBQ = 30°$. Find the measure of the angle $\angle AQB$.

Solution

The solution presented below refers to Figure 1.1.

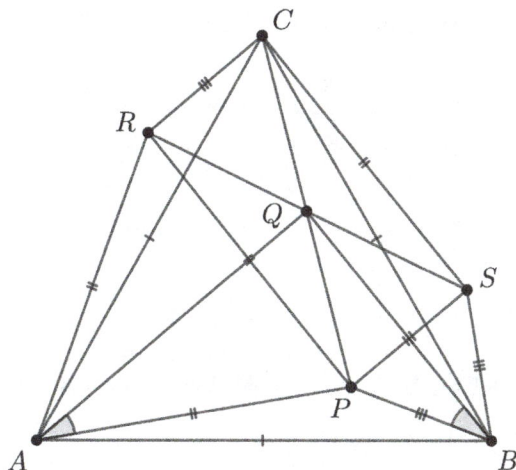

Figure 1.1   Rotation with center A by $60°$ in Problem 3.

Answer: $\angle AQB = 90°$.

Let us consider a rotation with center A by $60°$ that takes the point B to the point C, and the point P to some other point R. The triangles ABP and ACR are congruent and $RC = BP$. The triangle APR is equilateral and $AP = PR = RA$.

Let us consider a rotation with center B by $60°$ that takes the point A to the point C, and the point P to some other point S. The triangles ABP and CBS are congruent and $CS = AP$. The triangle BPS is equilateral and $BP = PS = SB$.

Since $\angle PAQ = 30°$, then $\angle RAQ = 30°$ and AQ is the perpendicular bisector to the segment PQ. Therefore $QR = QP$.

Similarly BQ is the perpendicular bisector to the segment PS and therefore $QP = QS$. Hence

$$QP = QS = SC = SQ = r$$

and the points R, C, S and P lie on the circle with center Q and radius r.

Since $CS = AP$ and $RC = BP$, then the quadrilateral $CRPS$ is a rectangle and $\angle RPS = 90°$ and thus $\angle AQB = 90°$.

Problem 4 (Number Theory)

Find all triples of integer numbers (x, y, z), such that

$$x^2 + y^2 = 3z^2$$

Solution

Answer: $(0, 0, 0)$.

Start by noticing that the triple $(0, 0, 0)$ satisfies the equation as

$$(0)^2 + (0)^2 = 3(0)^2$$

From now on we will assume the triple of numbers (x, y, z) is such that not all three of the numbers are zero. Let d be their greatest common divisor and

$$x = dx_1$$
$$y = dy_1$$
$$z = dz_1$$

where $\gcd(x_1, y_1, z_1) = 1$.

Now the equation becomes

$$x^2 + y^2 = 3z^2$$
$$(dx_1)^2 + (dy_1)^2 = 3(dz_1)^2$$
$$d^2 x_1^2 + d^2 y_1^2 = 3d^2 z_1^2$$
$$x_1^2 + y_1^2 = 3z_1^2$$

Notice that if n is not divisible by 3, then $n^2 \equiv 1 \pmod 3$.

We will proceed by doing the following casework:

- If x_1 and y_1 are not divisible by 3, then

$$x_1^2 + y_1^2 \equiv 2 \pmod 3$$

which contradicts the fact that it should be divisible by 3.

- If only one of x_1 or y_1 is divisible by 3, the other number should also be divisible by 3, which leads to a contradiction.

- If both x_1 and y_1 are divisible by 3, then $x_1^2 + y_1^2$ is divisible by 9 and, therefore, z_1 should be divisible by 3, which contradicts the fact that $\gcd(x_1, y_1, z_1) = 1$.

Problem 5 (Algebra)

Prove that if $a, b, c > 0$, then

$$\frac{a^3}{bc} + \frac{b^3}{ca} + \frac{c^3}{ab} \geq a + b + c$$

Solution

Let us apply the Chebyshev's Inequality to the numbers (a^3, b^3, c^3) and $(\frac{1}{bc}, \frac{1}{ca}, \frac{1}{ab})$:

$$\frac{a^3}{bc} + \frac{b^3}{ca} + \frac{c^3}{ab} \geq \frac{1}{3}(a^3 + b^3 + c^3)\left(\frac{1}{bc} + \frac{1}{ca} + \frac{1}{ab}\right)$$

However, by Titu's Lemma we have

$$\frac{1}{bc} + \frac{1}{ca} + \frac{1}{ab} \geq \frac{(1+1+1)^2}{ab + bc + ca} = \frac{9}{ab + bc + ca}$$

Therefore, we have shown that

$$\frac{a^3}{bc} + \frac{b^3}{ca} + \frac{c^3}{ab} \geq \frac{3\left(a^3 + b^3 + c^3\right)}{ab + bc + ca}$$

Let us apply the Chebyshev's Inequality to the numbers (a, b, c) and $\left(a^2, b^2, c^2\right)$:

$$a^3 + b^3 + c^3 \geq \frac{1}{3}(a + b + c)\left(a^2 + b^2 + c^2\right)$$

Since $a^2 + b^2 + c^2 \geq ab + bc + ca$, we have

$$\frac{a^3}{bc} + \frac{b^3}{ca} + \frac{c^3}{ab} \geq \frac{(a + b + c)\left(a^2 + b^2 + c^2\right)}{ab + bc + ca} \geq a + b + c$$

which is what needed to be proven.

Problem 6 (Combinatorics)

Let x_1, x_2, x_3, ... be a sequence of positive integers such that, for every positive integers m, n:

$$x_{mn} \neq x_{m(n+1)}$$

Prove that there exists a positive integer i such that $x_i \geq 2017$.

Solution

Let us call two numbers i and j *connected* if one of them can be written in the form mn and the other in the form $m(n + 1)$ for some $m, n \in \mathbb{N}$. This means that x_i is different from x_j whenever i and j are *connected*.

We want to show that there are at least 2017 positive integers i_1, ... , i_{2017}, such that 2017 terms of the sequence x_{i_1}, ... , $x_{i_{2017}}$ are all different from each other. We just need to demonstrate that there are 2017 positive integers i_1, ... , i_{2017} that are pairwise *connected*. We will prove this by induction on n.

Claim

For every $n \in \mathbb{N}$ there exist n positive integers i_1, ... , i_n that are pairwise *connected*.

Proof

For the basis of the induction, $n = 1$, we can just set $1 = 1$.

For inductive step, we observe that i and j, with $i < j$, are *connected* if and only if $j - i$ is a divisor of i. So if i and j are *connected* and k is a multiple of i then $k + i$ and $k + j$ are *connected*.

Let us assume that n positive integers $i_1, \ldots, i_n$ are pairwise *connected*. Let t be their product. Therefore, the numbers $t + i_1, \ldots, t + i_n$ are pairwise *connected*. These numbers are also all *connected* to k. Thus we constructed a set of $n + 1$ pairwise connected numbers. ■

CHAPTER 2

SENIOR LEVEL EXAM OF 2018

Problem 1 (Combinatorics)

100 nonzero numbers are written around the circle. For any two consecutive numbers we write between them their product. After this is done we erase all the old numbers. It is known that the number of positive numbers around the circle did not change. What is the smallest number of positive numbers that can be initially written around the circle?

Solution

Answer: 34.

Let us assume that there are at most 33 positive numbers around the circle.

Notice that the negative numbers can be formed only if there was exactly one positive number in the pair. Every positive number can only participate in the creation of 2 such negative numbers. Therefore, the number f negative numbers is at most

$$2 \cdot 33 = 66$$

This implies that the total number of numbers is at most 99 and we obtained a contradiction.

Junior, Intermediate and Senior Math Olympiads
by Roman Kvasov, Ph.D.

133

It is not hard to see that the case of 34 positive and 66 negative numbers can be arranged in the following way:

$$+ , - , - , + , - , - , \ldots$$

Problem 2 (Geometry)

Given a quadrilateral $ABCD$, such that no two of its sides are parallel. Let P be an arbitrary point on the side AD. Let Q be the second point of intersection of the circumcircles of the triangles ABP and CDP. Show that the line PQ passes through a fixed point that does not depend on the position of the point P on the segment AD.

Solution

The solution presented below refers to Figure 2.1.

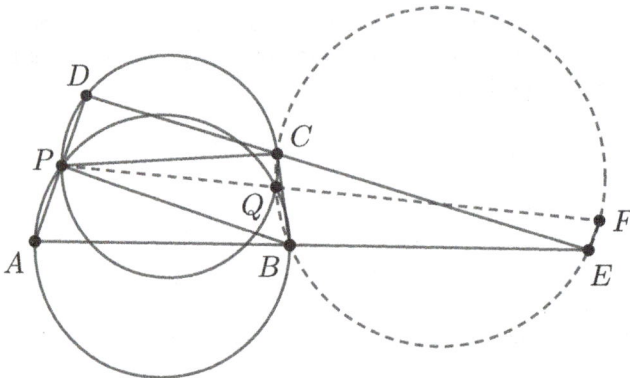

Figure 2.1   Quadrilaterals $CQPD$ and $BQPA$ are cyclic in Problem 2.

Let E be the intersection of the lines AB and CD. Let ω be the circumcenter of the triangle BCE.

The quadrilaterals $CQPD$ and $BQPA$ are cyclic and therefore $\angle CQP = \angle EDP$ and $\angle PQB = \angle PAE$ and $CQBE$ is also cyclic.

Let F be the second intersection of the line PQ with ω. Then $QCEF$ is cyclic and $PD \parallel EF$. This implies that F is the needed point.

Problem 3 (Number Theory)

Given a prime number $p > 2$ and positive integers m and n. It is known that p divides the numbers $m^2 + n^2$ and $m^3 + n^3$. Prove that p must divide the numbers m and n.

Solution

Notice that since p divides the numbers $m^2 + n^2$, then

$$m^2 \equiv -n^2 \pmod{p}$$

Let us now factor the expression $m^3 + n^3$:

$$m^3 + n^3 = (m + n)\left(m^2 - mn + n^2\right)$$

Since p divides the numbers $m^3 + n^3$, then we have that p divides $m + n$ or p divides $m^2 - mn + n^2$. We will proceed by doing the following casework.

- If p divides $m + n$, then

$$m \equiv -n \pmod{p}$$

  Therefore

$$m^2 + n^2 \equiv 0 \pmod{p}$$
$$(-n)^2 + n^2 \equiv 0 \pmod{p}$$
$$2n^2 \equiv 0 \pmod{p}$$

  Since p is prime and $p > 2$, then p divides n. Since p divides $m + n$, then p also divides m, as desired.

- If p divides $m^2 - mn + n^2$, then taking into account that

$$m^2 + n^2 \equiv 0 \pmod{p}$$

  we have

$$m^2 - mn + n^2 \equiv 0 \pmod{p}$$
$$-mn \equiv 0 \pmod{p}$$
$$mn \equiv 0 \pmod{p}$$

  Since p is prime, then p divides m or p divides n. Since p divides $m + n$, then p divides both numbers m and n, as desired.

Problem 4 (Geometry)

Let O be the circumcenter of an acute triangle ABC. Line k is perpendicular to AC and intersects the side BC at Q and the extension of the side AB at P. Show that the midpoints of the segments AP and CQ and the points O and B lie on the same circle.

Solution

The solution presented below refers to Figure 2.2.

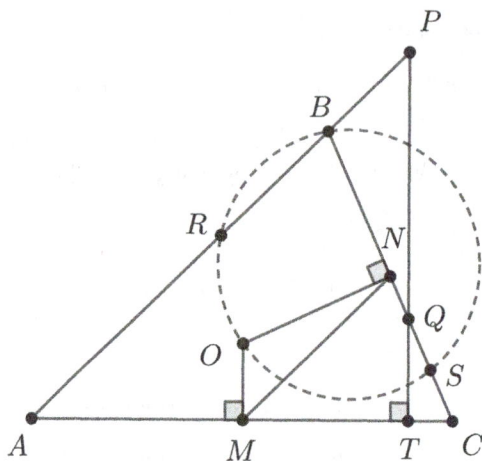

Figure 2.2   Triangles MON and PBQ are similar in Problem 4.

Let R, S, M, N be the midpoints of the segments AP, CQ, AC and BC, respectively. It will be enough to prove that the quadrilateral $ORBS$ is cyclic.

Notice that we have $OM \perp AB$ and $MN \parallel AC$. From here also

$$\angle OMN = 90° - \angle NMB = 90° - \angle BAC = \angle BPQ$$

Also

$$\angle MON = 180° - \angle ABC = \angle PBQ$$

Therefore, the triangles MON and PBQ are similar and

$$\frac{OM}{BQ} = \frac{ON}{BP}$$

However, since $BP = 2MR$ and $BQ = 2NS$, then

$$\frac{OM}{ON} = \frac{BQ}{BP} = \frac{NS}{MR}$$

This implies that the triangles MRO and NSO are similar. From here we have that $\angle MRO = \angle NSO$ and the quadrilateral $ORBS$ is cyclic, as desired.

Problem 5 (Number Theory)

Find all pairs of natural numbers a and k, such that for any natural n relatively prime with a, the number $a^{k^n+1} - 1$ is a multiple of n.

Solution

Answer: $a = 1$ and k is any natural number.

Let us do the following casework:

- If $a = 1$, then
$$a^{k^n+1} - 1 = 1 - 1 = 0$$

  which is divisible by n for any $k \in \mathbb{N}$.

- If $a \geq 2$, then let us put $n = a^k - 1$. Notice that
$$\gcd(n, a) = \gcd\left(a^k - 1, a\right) = 1$$

  We have that
$$a^k - 1 \equiv 0 \pmod{n}$$

  or equivalently
$$a^k \equiv 1 \pmod{n}$$

  Therefore
$$0 \equiv a^{k^n+1} - 1 \equiv a^{k^n} \cdot a - 1 \equiv a - 1 \pmod{n}$$

  For $k = 1$ we have that
$$a^{k^n+1} - 1 = a^2 - 1$$

  and should be divisible by all n relatively prime with a, which is impossible.

For $k > 1$ we have that

$$0 < a - 1 < a^k - 1$$

and, therefore, cannot be divisible by $a^k - 1$.

We conclude that $a = 1$ and k is any natural number.

Problem 6 (Algebra)

Let a, b, c, d be distinct real numbers, such that $ac = bd$. Given that

$$\frac{a}{b} + \frac{b}{c} + \frac{c}{d} + \frac{d}{a} = 4$$

Find the maximum value of the expression

$$M = \frac{a}{c} + \frac{b}{d} + \frac{c}{a} + \frac{d}{b}$$

Solution

Answer: the maximum value of M is -12.

Let us make the following substitutions:

$$\frac{a}{b} = x$$

$$\frac{b}{c} = y$$

Notice that since a, b, c, d are distinct, then $x \neq 1$ and $y \neq 1$. Furthermore, we have

$$\frac{a}{c} = xy$$

$$\frac{c}{a} = \frac{1}{xy}$$

$$\frac{c}{d} = \frac{b}{a} = \frac{1}{x}$$

$$\frac{d}{a} = \frac{c}{b} = \frac{1}{y}$$

Also

$$\frac{b}{d} = \frac{a}{d} \cdot \frac{c}{d} = \frac{y}{x}$$

and

$$\frac{d}{b} = \frac{x}{y}$$

Therefore, the problem can be rewritten as follows: find the maximum of

$$M = xy + \frac{x}{y} + \frac{1}{xy} + \frac{y}{x}$$

given that

$$\left(x + \frac{1}{x}\right) + \left(y + \frac{1}{y}\right) = 4$$

Notice that from the last equality x and y cannot be both positive. Indeed, for positive x and y by the AM-GM Inequality:

$$4 = \left(x + \frac{1}{x}\right) + \left(y + \frac{1}{y}\right) \geq (2) + (2) = 4$$

However, this means that in the AM-GM Inequality the equality holds, which in turn implies that $x = y = 1$ and leads to a contradiction.

Therefore, at least one of x or y is negative. Without loss of generality let it be x.

Let us factor the expression for M:

$$M = xy + \frac{x}{y} + \frac{1}{xy} + \frac{y}{x} = x\left(y + \frac{1}{y}\right) + \frac{1}{x}\left(\frac{1}{y} + y\right) = \left(x + \frac{1}{x}\right)\left(y + \frac{1}{y}\right)$$

If we put $x + \frac{1}{x} = t$, then $y + \frac{1}{y} = 4 - t$. Then the expression for M becomes

$$M = t(4 - t) = -t^2 + 4t$$

Notice that since x is negative, then by the AM-GM Inequality applied to the numbers $-x$ and $\frac{1}{-x}$ we have

$$-x + \frac{1}{-x} \geq 2$$

which implies that

$$x + \frac{1}{x} \leq -2$$

and, therefore, $t \leq -2$.

Let us now consider the quadratic function $f(t) = -t^2 + 4t$. The maximum of the function $f(t)$ is located at $t = 2$, and the function is increasing on the interval $(-\infty, 2]$. However, since in our case $t \leq -2$, then the maximum of the function $f(t)$ on the interval $(-\infty, -2]$ is reached at $t = -2$, which in turn is reached for $x = -1$ and $a = -b$. From here we have

$$M = f(t) \leq f(-2) = -(-2)^2 + 4(-2) = -12$$

as desired.

CHAPTER 3

SENIOR LEVEL EXAM OF 2019

Problem 1 (Combinatorics)

Given an infinite square paper and a set N of $4n$ randomly chosen squares. Is it always possible to choose at least n squares from the set N, such that no two of the chosen squares share a common point?

Solution

Answer: yes, it is possible.

We will paint all the squares of the plane in 4 colors (see Figure 3.1).

Figure 3.1   Squares are painted in four colors in Problem 1.

Junior, Intermediate and Senior Math Olympiads
by Roman Kvasov, Ph.D.

141

Notice that the squares of the same color do not have a common point. Now by the Pigeonhole Principle at least n squares from the set N are of the same color. Since these squares do not share a point, we can choose them to be the needed n squares.

Problem 2 (Geometry)

Let ABC be an acute triangle, with $AB \neq AC$ and centroid G. Let M be the midpoint of BC. Consider the circumference Γ with center G and radius GM and denote by N the intersection of Γ with BC different from M. Let S be the point symmetric to A with respect to N. Prove that $GS \perp BC$.

Solution

The solution presented below refers to Figure 3.2.

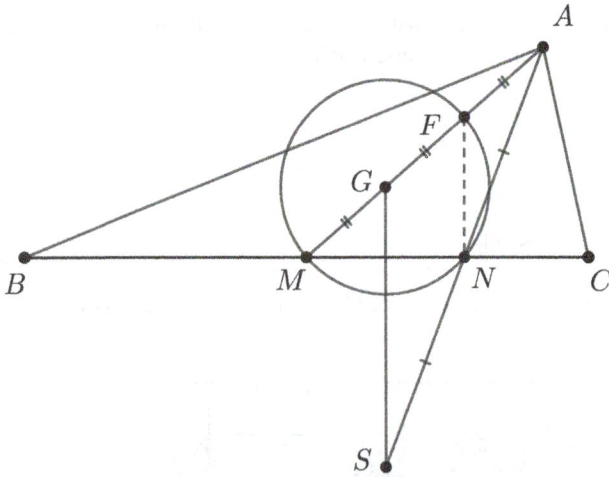

Figure 3.2 FN is the midsegment of the triangle AGS in Problem 3.

Let F be the midpoint of AG. Notice that since the centroid divides the medians in the ratio $2:1$, then F belongs to Γ and FM is the diameter of Γ.

Since N is the midpoint of AS, then FN is the midsegment of the triangle AGS. This implies that $FN \parallel GS$.

Since FM is the diameter of Γ, then $\angle MNF = 90°$ and $FN \perp BC$. From here $GS \perp BC$, which is what needed to be proven.

Problem 3 (Number Theory)

Let m and n be positive integers and p a prime number for which $m < n < p$. Assume that p divides the numbers $m^2 + 1$ and $n^2 + 1$. Prove that p also divides the number $mn - 1$.

Solution

Since p divides the numbers $m^2 + 1$ and $n^2 + 1$, the it also divides their difference:

$$\left(n^2 + 1\right) - \left(m^2 + 1\right) = n^2 + 1 - m^2 - 1 = n^2 - m^2 = (n - m)(n + m)$$

Since p is a prime, then p should divide $n - m$ or $n + m$. We will proceed by doing the following casework:

- If p divides $n - m$, then we have

$$0 < n - m < n < p$$

  which contradicts the fact that $n - m$ is a multiple of p.

- If p divides $n + m$, then we have

$$n + m \equiv 0 \pmod{p}$$
$$n \equiv -m \pmod{p}$$

  Therefore

$$mn - 1 \equiv m(-m) - 1 \equiv -m^2 - 1 \equiv -\left(m^2 + 1\right) \equiv 0 \pmod{p}$$

  which implies that p divides the number $mn - 1$, as desired.

Problem 4 (Geometry)

Let A and B be the points of intersection of two circles. A line passing through the point A intersects the circles at the points C and D. Let P and Q be the projections of the point B onto the tangents to the circles at the points C and D, respectively. Show that PQ is tangent to the circle with diameter AB.

Solution

The solution presented below refers to Figure 3.3.

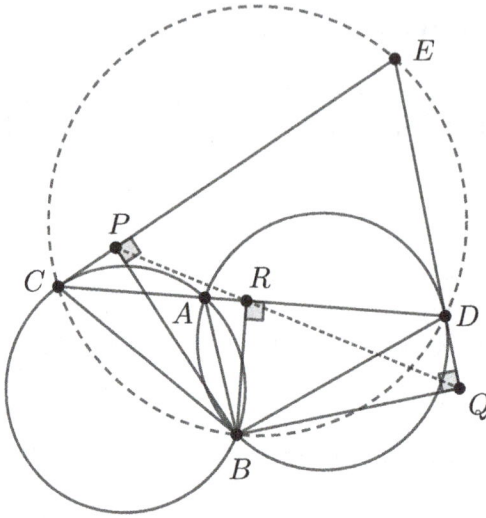

Figure 3.3   Points P, Q and R lie on the Simson line in Problem 4.

Let R be the projection of the point B onto the line CD. Since $BR \perp AR$, then R lies on the circle with diameter AB.

Let us put $\angle ABC = \alpha$ and $\angle ABD = \beta$. Then

$$\angle CBD = \angle ABC + \angle ABD = \alpha + \beta$$

Using the theorem about the tangent and the chord we have:

$$\angle CED = 180° - \angle ECD - \angle EDC$$
$$= 180° - \angle ECA - \angle EDA$$
$$= 180° - \angle ABC - \angle ABD$$
$$= 180° - \alpha - \beta$$

From here we see that

$$\angle CBD + \angle CED = 180°$$

and $CEDB$ is cyclic, or equivalently, B lies on the circumcircle of the triangle CED. Since P, Q and R are the projections of the point B onto the lines CE, ED and CD, then the points P, Q and R lie on the Simson line.

We will not prove that the line PQ is tangent to the circle with diameter AB at the point R. It will be enough to prove that $\angle ABR = \angle PRC$. Indeed, let us put $\angle PRC = \gamma$. Then $\angle PCB = 90° - \gamma$. From the theorem about the tangent and the chord we have that $\angle CAB = 90° + \gamma$ and, therefore, $\angle BAR = 90° - \gamma$. From here $\angle ABR = \gamma$, as desired.

Problem 5 (Algebra)

The plane is tiled with congruent equilateral triangles of side 1. Let d_1 be the distance between some vertices A_1 and B_1 and d_2 be the distance between some vertices A_2 and B_2. Is it possible to find the vertices A_3 and B_3, such that the distance d_3 between A_3 and B_3 is equal to $d_1 \cdot d_2$?

Solution

Answer: yes, it is possible.

Let us consider two unit vectors e_1 and e_2 that form $120°$ and are collinear with the sides of the triangles.

Let us put

$$\overrightarrow{A_1B_1} = ae_1 + be_2$$

$$\overrightarrow{A_2B_2} = ce_1 + de_2$$

for some integers a, b, c, d. Then we have

$$d_1^2 = \overrightarrow{A_1B_1} \cdot \overrightarrow{A_1B_1} = a^2 - ab + b^2$$

$$d_2^2 = \overrightarrow{A_2B_2} \cdot \overrightarrow{A_2B_2} = c^2 - cd + d^2$$

Now let us consider the following product:

$$d_1^2 \cdot d_2^2 = \left(a^2 - ab + b^2\right)\left(c^2 - cd + d^2\right) = x^2 - xy + y^2$$

where $x = ac - bd$ and $y = ad + bc - db$.

From here we conclude that the vector

$$\overrightarrow{A_3B_3} = xe_1 + ye_2$$

satisfies the conditions of the problem.

Problem 6 (Number Theory)

Show that there exists an infinite number of positive integers of the form 5^n, such that their decimal representation contains at least 2019 consecutive zeros.

Solution

First let us show that for any $k \in \mathbb{N}$, there exist infinitely many $m \in \mathbb{N}$, such that

$$5^m \equiv 1 \pmod{2^k}$$

Let us consider the sequence

$$5^0, 5^1, 5^2, \ldots, 5^{2^k}$$

By Pigeonhole principle there are two numbers 5^p and 5^q ($p > q$) that have the same remainders modulo 2^k. Therefore, 2^k divides the number

$$5^p - 5^q = 5^q(5^{p-q} - 1)$$

and then

$$5^{p-q} \equiv 1 \pmod{2^k}$$

Since

$$\left(5^{p-q}\right)^i \equiv 1^i \equiv 1 \pmod{2^k}$$

then for $m = (p-q)i$, $i \in \mathbb{N}$, we have

$$5^m \equiv 1 \pmod{2^k}$$

Therefore

$$5^{m+k} \equiv 5^k \pmod{10^k}$$

which means that the last k digits of the number 5^{m+k} are exactly the same as the number 5^k with possibly some zeros in front of it. For $2^k > 10^{2019}$ we have that

$$5^k = \frac{10^k}{2^k} < \frac{10^k}{10^{2019}} = 10^{k-2019}$$

has no more than $k - 2019$ digits. Therefore, from the last k digits of the number 5^{m+k} only the last $k - 2019$ digits are non-zero. This implies that the rest 2019 digits should be zero digits.

CHAPTER 4

SENIOR LEVEL EXAM OF 2020

Problem 1 (Geometry)

Let ABC be an isosceles triangle with $AB = AC$ and let D and E be the points on the sides AB, BC respectively, such that the lines $DE \parallel AC$. Let the point F be on the line DE, which is on the opposite side of D with respect to E and is such that $FE = AD$. Let O be the circumcenter of the triangle BDE. Prove that the points O, F, A, D lie on the same circle.

Solution

The solution presented below refers to Figure 4.1.

Notice that since O is the circumcenter of the triangle BDE, then $OD = OE$ and $\angle ODE = \angle OED$.

Since the triangle ABC is isosceles and $DE \parallel AC$, then the triangle BDE is also isosceles. Therefore DO is the angle bisector and $\angle ODE = \angle ODB$.

We also have that

$$\angle ODA = 180° - \angle ODB = 180° - \angle OED = \angle OEF$$

This implies that the triangles ODA and OEF are congruent by the Side-Angle-Side Postulate.

Junior, Intermediate and Senior Math Olympiads
by Roman Kvasov, Ph.D.

147

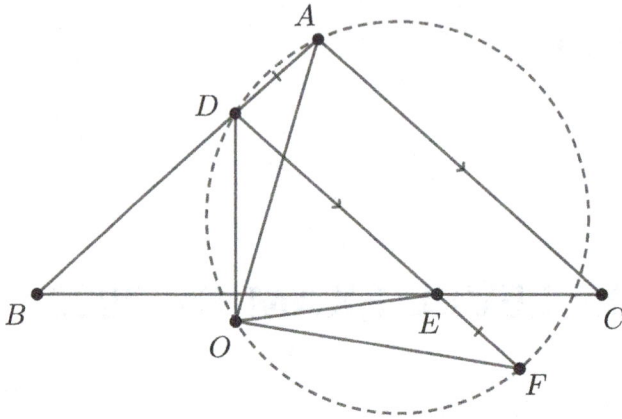

Figure 4.1   Triangles ODA and OEF are congruent in Problem 1.

From here

$$\angle DAO = \angle EFO$$

and the points O, F, A, D lie on the same circle, as desired.

Problem 2 (Number Theory)

Let p_1, p_2, $\ldots$, p_n be distinct prime numbers and P be their product. Let the number S be defined as

$$S = \sum_{i=1}^{n} \left(\frac{P}{p_i} \right)^{p_i - 1}$$

Show that $S - 1$ is a multiple of P.

Solution

Start by noticing that $\frac{P}{p_i}$ is an integer and

$$\gcd\left(\frac{P}{p_i}, p_i \right) = 1$$

for all $i = 1, 2, \ldots, n$.

Now by Fermat's Little Theorem we have

$$\left(\frac{P}{p_i}\right)^{p_i-1} \equiv 1 \pmod{p_i}$$

Notice also that $\frac{P}{p_i}$ is divisible by p_j for all $i \neq j$. Therefore

$$S \equiv 1 \pmod{p_i}$$

and $p_i \mid (S-1)$ for all $i = 1, 2, \ldots, n$.

Since p_i are all distinct, then $S - 1$ is divisible by their product and $S - 1$ is a multiple of P.

Problem 3 (Combinatorics)

Ariana wrote the first 10001 terms of the Fibonacci sequence on the board

$$0, 1, 1, 2, 3, 5, 8, 13, \ldots$$

where each next term equals to the sum of the two previous terms. Is it true that among the numbers on the board there is a number that ends with two zeros?

Solution

Answer: Yes, it is true.

Let a_1, a_2, a_3, ... be the sequence of remainders modulo 100 of the given Fibonacci sequence, i.e. $a_1 = 0$, $a_2 = 1$ and

$$a_{n+2} \equiv a_{n+1} + a_n \pmod{100}$$

and $0 \leq a_n \leq 99$ for all $n \in \mathbb{N}$.

Notice that $a_1 = 0$ and it will be enough to prove that among the first 10001 terms of the sequence a_n there is another zero term.

We will first prove that the sequence a_n is periodic with a period of length less than 10001. Let us consider the 10000 pairs of numbers:

$$(a_1, a_2), (a_2, a_3), \ldots, (a_{10000}, a_{10001})$$

Since there is no pair $(0, 0)$, then by the Pigeonhole Principle there exist two pairs that are exactly the same. Since the value of a_{n+2} is completely determined by the pair (a_n, a_{n+1}), then the sequence a_n is periodic with a period less than 10001. Since $a_1 = 0$, then there exists the positive integer $i \leq 10001$, such that $a_i = 0$ as desired.

Problem 4 (Algebra)

Prove the inequality for positive numbers

$$\frac{a^6 + 1}{b} + \frac{b^6 + 1}{a} \geq 2 \left(a^2 + b^2 \right)$$

Solution

Let us start by applying the AM-GM Inequality to the numerators of both fractions:

$$a^6 + 1 \geq 2\sqrt{a^6} = 2a^3$$
$$b^6 + 1 \geq 2\sqrt{b^6} = 2b^3$$

Now it will be enough to prove that

$$\frac{2a^3}{b} + \frac{2b^3}{a} \geq 2 \left(a^2 + b^2 \right)$$

which is equivalent to

$$\frac{a^3}{b} + \frac{b^3}{a} \geq a^2 + b^2$$
$$\frac{a^4 + b^4}{ab} \geq a^2 + b^2$$
$$a^4 + b^4 \geq a^3 b + a b^3$$
$$a^4 + b^4 - a^3 b - a b^3 \geq 0$$
$$a^3 (a - b) - b^3 (a - b) \geq 0$$
$$(a - b) \left(a^3 - b^3 \right) \geq 0$$
$$(a - b)(a - b) \left(a^2 + ab + b^2 \right) \geq 0$$
$$(a - b)^2 \left(a^2 + ab + b^2 \right) \geq 0$$

which obviously holds.

Problem 5 (Combinatorics)

A positive integer number n is written on the board. It is allowed to take any written number m, erase it and then write two positive integer numbers x and y, such that $xy = 2m^2$. Prove that if at some moment there are 100 numbers written on the board, then one of them is not greater than $10n$.

Solution

Let x_1, x_2, $\ldots$, x_m be the numbers on the board after $m - 1$ operations.

Consider the sums of squares of the reciprocals of these numbers:

$$S_m = \sum_{i=1}^{m} \left(\frac{1}{x_i^2} \right)$$

From $xy = 2n^2$ we have

$$\frac{1}{2n^2} = \frac{1}{xy} \leq \frac{1}{2} \left(\frac{1}{x^2} + \frac{1}{y^2} \right)$$

by AM-GM inequality.

From here

$$\frac{1}{n^2} \leq \frac{1}{x^2} + \frac{1}{y^2}$$

and, therefore, we have

$$S_1 \leq S_2 \leq \ldots \leq S_m$$

for all $m \in \mathbb{N}$.

Let z be the smallest number on the board after 99 operations, i.e. $z \leq x_i$ for all $i = 1, 2, \ldots, 100$. Then we have

$$\frac{1}{n^2} = S_1 \leq S_2 \leq \ldots \leq S_{100} \leq 100 \cdot \frac{1}{z^2}$$

which implies that

$$z \leq 10n$$

Problem 6 (Geometry)

Let $ABCD$ be a parallelogram and P be the point on the segment CD. Let Q be the intersection of the lines AP and BC. Let R be the point on the circumcircle of the triangle BCD, such that $RP = RQ = RC$. Prove that the line AP bisects the angle $\angle BAD$.

Solution

The solution presented below refers to Figure 4.2.

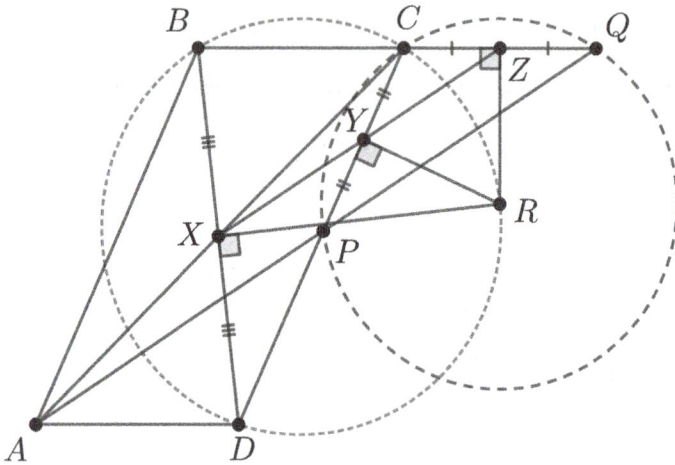

Figure 4.2   Homothecy with center at C and coefficient $1/2$ takes the points A, P and Q to the points X, Y and Z in Problem 6.

Notice that since $RP = RQ = RC$, then R lies on the perpendicular bisectors to the segments PC, CQ and PQ. This implies that R is circumcenter of the triangle PCQ.

Let X, Y and Z be the midpoints of the segments BD, PC and CQ, respectively. Notice that X is the point intersection of the diagonals of $ABCD$ and $AX = XC$. If we consider homothecy with center at C and coefficient $1/2$, then the points A, P and Q will become the points X, Y and Z, respectively. Since A, P and Q, then so are X, Y and Z.

Since Y and Z are the midpoints of PC and CQ, then $RY \perp PC$ and $RZ \perp CQ$. Since R is on the circumcircle of BCD, then Y and Z are the feet of the perpendiculars from the point R to the lines CD and BC. This implies that the line YZ is the Simson line. Moreover, since YZ intersects the line BD at X, then X is the foot of the perpendicular from the point R to the lines BD. From here $RX \perp BD$ and the triangle BRD is isosceles.

Let us put

$$\angle RBD = \angle RDB = \alpha$$

Since $BCRD$ is cyclic, then $\angle RCD = \alpha$ and $\angle RCQ = \alpha$.
From here

$$\angle PRC = \angle CRQ = 180° - 2\alpha$$

Since

$$\angle PAD = \angle PQC = \frac{1}{2}\angle PRC = 90° - \alpha$$

and

$$\angle PAB = \angle APD = \angle CPQ = \frac{1}{2}\angle CRQ = 90° - \alpha$$

then $\angle PAD = \angle PAB$ and AP bisects the angle $\angle BAD$, as desired.

CHAPTER 5

SENIOR LEVEL EXAM OF 2021

Problem 1 (Geometry)

Let ABC be a right triangle with $\angle ACB = 90°$ and let K, L and M be the points of tangency of the sides AC, BC and AB with its incircle. Let H be the orthocenter of the triangle KLM. Prove that $CH \perp AB$.

Solution

The solution presented below refers to Figure 5.1.

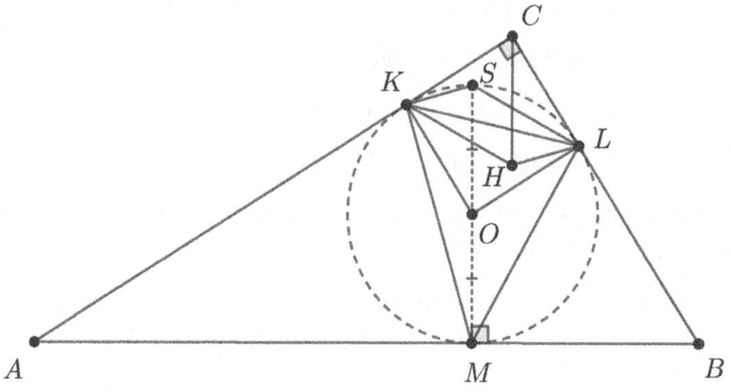

Figure 5.1 $SKHL$ is a parallelogram and $CKOL$ is a square in Problem 1.

Junior, Intermediate and Senior Math Olympiads
by Roman Kvasov, Ph.D.

155

Let O be the center of the incircle and MS be the diameter of the incircle.

Then SK and LH are parallel as perpendiculars to MK. Similarly, SL and KH are parallel, and therefore $SKHL$ is a parallelogram.

Since $CKOL$ is a square, the segments CH and OS are symmetric about the midpoint of the segment KL. Therefore $CH \parallel MS$ and $CH \perp AB$.

Problem 2 (Combinatorics)

A positive integer is written on each face of a cube. Each vertex of the cube is assigned the multiplication of the numbers of the three faces that have that vertex in common. The sum of the 8 numbers assigned to the vertices is 114. Determine the sum S of the numbers on the faces.

Solution

Answer: 24.

Let a, b, c, d, e and f be the numbers on the faces of the cube, such that the faces with the numbers a and d are parallel, b and e are parallel, c and f are parallel.

The sum of the numbers at the vertices is

$$114 = abc + abf + ace + aef + bcd + bdf + cde + def$$
$$= ab(c + f) + ae(c + f) + bd(c + f) + de(c + f)$$
$$= (c + f)(ab + ae + bd + de)$$
$$= (c + f)(a(b + e) + d(b + e))$$
$$= (c + f)(b + e)(a + d)$$

This implies that the numbers $(a + d)$, $(b + e)$, $(c + f)$ are the factors of 114. Since a, b, c, d, e and f are positive integers, then

$$(a + d) \geq 2$$
$$(b + e) \geq 2$$
$$(c + f) \geq 2$$

Notice that since the prime factorization of the number 114 is

$$114 = 2 \cdot 3 \cdot 19$$

then the values of the quantities $(a+d)$, $(b+e)$ and $(c+f)$ are exactly 2, 3 and 19 in some order. From here the sum of the numbers on the faces is equal to

$$\begin{aligned} S &= a+b+c+d+e+f \\ &= (a+d) + (b+e) + (c+f) \\ &= (2) + (3) + (19) \\ &= 24 \end{aligned}$$

as desired.

Problem 3 (Number Theory)

Prove that there exists an infinite number of positive integers that cannot be represented in the form

$$x^{yz} - y^{xw}$$

where x, y, z and w are positive integers with $x, y > 1$.

Solution

We will work modulo 8 and show that no number of the form $8n + 3$ can be represented in this way.

Indeed, suppose that

$$x^{yz} - y^{xw} = 8n + 3$$

for some $n \in \mathbb{N}$.

We will proceed by doing the following casework:

- If the numbers x and y have the same parity, then the number $x^{yz} - y^{xw}$ is even. We obtained a contradiction.

- If one of the numbers x and y is even, and the other is odd, then the odd number is greater of equal to 3. From here, one of the numbers x^{yz} or y^{xw} is divisible by 8, and the other is a square of an odd integer, and therefore, is congruent to 1 modulo 8. Consequently, the difference of the numbers cannot be congruent to 3 modulo 8. We obtained a contradiction.

We conclude that there exists an infinite number of positive integers that cannot be represented in the given form, as desired.

Problem 4 (Combinatorics)

For some positive integer number n, consider an $n \times n$ board. On this board, arbitrary rectangles can be placed, with sides aligned along the grid lines. What is the smallest number of such rectangles needed so that all cells of the board are covered by a different number of rectangles (including the case of being covered by none of the rectangles)? The rectangles we place on the board can have the same sizes.

Solution

Answer: $n^2 - 1$.

If all n^2 cells are covered by a different number of rectangles, then some cell is covered by at least $n^2 - 1$ rectangles, and therefore, there are at least $n^2 - 1$ of them.

Now let us show that the needed configuration is possible for $n^2 - 1$ rectangles.

Let us denote the cell at the intersection of the j-th row and the k-th column as (j, k). For each $1 \leq j \leq n - 1$, place a rectangle on the board with the top left corner at $(1, 1)$ and the bottom right corner at (j, n). Also, for each $1 \leq j \leq n - 1$, place n rectangles with the top left corner at $(1, 1)$ and the bottom right corner at (n, j). The total number of rectangles, therefore, will be $n^2 - 1$.

The number of rectangles that cover the cell (j, k) is

$$(n - j) + n(n - k)$$

which are different for all $1 \leq j, k \leq n$, as desired.

Problem 5 (Number Theory)

Find all positive integer numbers n, for which there exists a prime number p, such that $3p + n^3$ is divisible by 33.

Solution

Answer: n is any multiple of 3.

We will solve the problem by doing the following casework:

- If n is not divisible by 3, then $3p + n^3$ is not divisible by 3 and therefore cannot be divisible by 33. Contradiction.

- If n is divisible by 3 and divisible by 11, then $n = 33k$ and we have that the number
$$\frac{3p + n^3}{33} = \frac{3p + (33k)^3}{33} = \frac{p}{11} + (33)^2 k^3$$
should be an integer. This implies that $p = 11$, which satisfies the conditions of the problem.

- If n is divisible by 3 and not divisible by 11, then $n = 3k$, where $\gcd(k, 11) = 1$. Then we have that the number
$$\frac{3p + n^3}{33} = \frac{3p + (3k)^3}{33} = \frac{p + 9k^3}{11}$$
should be an integer. By Dirichlet's Theorem, there exist infinitely many primes p of the form
$$p = 11m - 9k^3$$
Each of these primes satisfies the conditions of the problem.

We conclude that only those n that are multiples of 3 work.

Problem 6 (Algebra)

Given positive numbers a, b and c. Prove that

$$\sum_{\text{cyc}} \frac{2 + a^2}{\sqrt{1 + b^3}} \geq 6$$

Solution

Let us factor the expression under the root and apply AM-GM:

$$\sqrt{1 + b^3} = \sqrt{(1 + b)\left(1 - b + b^2\right)}$$
$$\leq \frac{(1 + b) + \left(1 - b + b^2\right)}{2}$$
$$= \frac{2 + b^2}{2}$$

From here we have

$$\sum_{cyc} \frac{2+a^2}{\sqrt{1+b^3}} \geq \sum_{cyc} \frac{2\left(2+a^2\right)}{2+b^2} = 2\sum_{cyc} \frac{2+a^2}{2+b^2}$$

However, notice that by AM-GM we have

$$\sum_{cyc} \frac{2+a^2}{2+b^2} \geq 3\sqrt[3]{\frac{2+a^2}{2+b^2} \cdot \frac{2+b^2}{2+c^2} \cdot \frac{2+c^2}{2+a^2}} = 3$$

Therefore, we have that

$$\sum_{cyc} \frac{2+a^2}{\sqrt{1+b^3}} \geq 2 \cdot 3 = 6$$

as desired.

CHAPTER 6

SENIOR LEVEL EXAM OF 2022

Problem 1 (Geometry)

In an acute triangle ABC the measure of the angle B is equal $60°$. The altitudes CE and AD intersect at the point H. Show that the circumcenter O of the triangle ABC lies on the common angle bisector of the angles $\angle AHE$ and $\angle CHD$.

Solution

The solution presented below refers to Figure 6.1.

Without loss of generality les us assume that the point O lies in the triangle AHE.

From the quadrilateral $BEHD$ we have that $\angle EHD = 120°$. From here we have that $\angle AHC = 120°$ and $\angle AHE = 60°$.

Since $\angle AOC$ is central, then it equals twice the angle $\angle ABC$, and therefore $\angle AOC = 120°$.

Since $\angle AOC = \angle AHC$, then the quadrilateral $AOHC$ is cyclic. Since the triangle AOC is isosceles and $\angle AOC = 120°$, then $\angle ACO = 30°$.

Junior, Intermediate and Senior Math Olympiads
by Roman Kvasov, Ph.D.

161

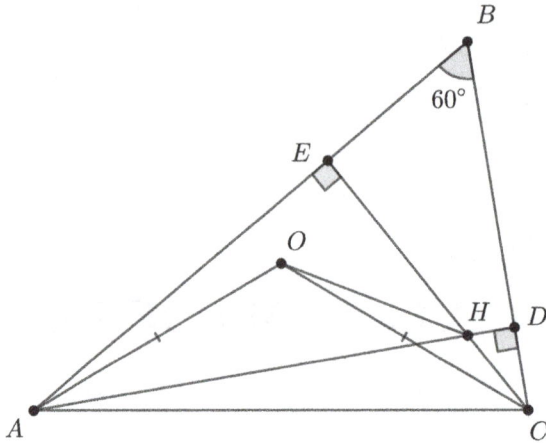

Figure 6.1   Quadrilateral $AOHC$ is cyclic in Problem 1.

Since $AOHC$ is cyclic, then $\angle AHO = \angle ACO = 30°$ and OH is the angle bisector of the angle $\angle AHE$, as desired.

Problem 2 (Algebra)

For the real numbers a, b, c and d from the interval $[0, 1]$, find the largest possible value of the expression

$$S = a^2 + b^2 + c^2 + d^2 - ab - bc - cd - ad$$

Solution

Answer: 2.

Let us start by noticing that the value of 2 is achieved for

$$a = 1, \quad b = 0, \quad c = 1, \quad d = 0$$

Let us now rewrite the expression for S as follows:

$$S = a^2 + b^2 + c^2 + d^2 - ab - bc - cd - ad$$
$$= \frac{(a-b)^2 + (b-c)^2 + (c-d)^2 + (d-a)^2}{2}$$

Since all variables belong to the interval $[0, 1]$, then each parenthesis squared does not exceed 1. Consequently, we have $S \leq 2$, as desired.

Problem 3 (Geometry)

In the cyclic quadrilateral $BCDE$, it is known that $\angle CBE = 2\angle BED$ and $DE = 2CD$. Let the point H be the projection of the point D onto the line BE. Prove that $CH \parallel DE$.

Solution

The solution presented below refers to Figure 6.2.

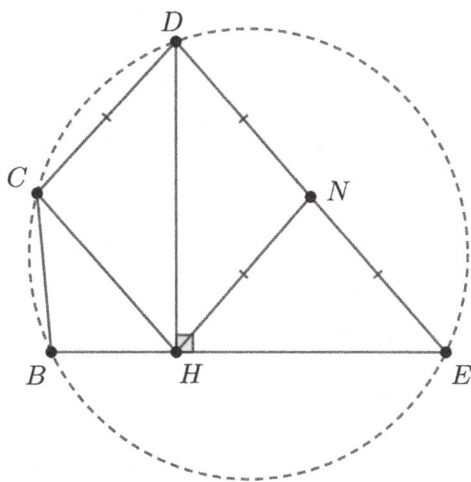

Figure 6.2 Quadrilateral $CDNH$ is a parallelogram in Problem 3.

Let M be the midpoint of the side DE. Since the median HN of the right triangle DHE is equal to half the hypotenuse, then

$$HN = EN = DN = CD$$

In addition, we have

$$\angle HNE = 180° - 2\angle BED = 180° - \angle CBE = \angle CDE$$

This implies that $CD \parallel NH$, and consequently, the quadrilateral $CDNH$ is a parallelogram. From here $CH \parallel DE$, as desired.

Problem 4 (Algebra)

Prove that if a, b, c are the lengths of the sides of some triangle, then

$$\frac{a}{b+c-a} + \frac{b}{a+c-b} + \frac{c}{a+b-c} \geq 3$$

Solution

Let us rewrite the LHS and apply Titu's Lemma:

$$\begin{aligned}
LHS &= \frac{a}{b+c-a} + \frac{b}{a+c-b} + \frac{c}{a+b-c} \\
&= \frac{a^2}{ab+ac-a^2} + \frac{b^2}{ab+bc-b^2} + \frac{c^2}{ac+bc-c^2} \\
&\geq \frac{(a+b+c)^2}{2ab+2bc+2ac-a^2-b^2-c^2} \\
&= \frac{a^2+b^2+c^2+2ab+2bc+2ac}{2ab+2bc+2ac-a^2-b^2-c^2}
\end{aligned}$$

Let us now show that

$$\frac{a^2+b^2+c^2+2ab+2bc+2ac}{2ab+2bc+2ac-a^2-b^2-c^2} \geq 3$$

Indeed, this inequality is equivalent to

$$\frac{a^2+b^2+c^2+2ab+2bc+2ac}{2ab+2bc+2ac-a^2-b^2-c^2} \geq 3$$
$$a^2+b^2+c^2+2ab+2bc+2ac \geq 6ab+6bc+6ac-3a^2-3b^2-3c^2$$
$$4a^2+4b^2+4c^2 \geq 4ab+4bc+4ac$$
$$2a^2+2b^2+2c^2 \geq 2ab+2bc+2ac$$
$$(a-b)^2+(b-c)^2+(c-a)^2 \geq 0$$

which obviously holds.

Problem 5 (Number Theory)

Find all positive integers n that satisfy the following two conditions: n is relatively prime to 42 and $20^n + 22^n$ is a multiple of n.

Solution

Answer: $n = 1$.

Start by noticing that $n = 1$ satisfies the conditions of the problem.

From now on we will assume that $n > 1$. Let us rewrite the expression as $2^n (10^n + 11^n)$ and let p be the smallest prime divisor of n.

We will proceed by doing the following casework:

- If $p = 2$, then n is not relatively prime with 42. We obtained a contradiction.

- If $p > 2$, then p is odd. Since n is relatively prime to 42, then p is different from 3 and 7. Notice that p divides $10^n + 11^n$ and, therefore, is relatively prime with 10 and 11. From here

$$10^n \equiv -11^n \pmod{p}$$

Squaring both sides we have

$$100^n \equiv 121^n \pmod{p}$$

Let x be such integer that

$$100x \equiv 121 \pmod{p}$$

From here we have that

$$100^n \equiv (100x)^n \pmod{p}$$

and, therefore

$$x^n \equiv 1 \pmod{p}$$

Let d be the order of the number x by modulo p. Then we have that $d \mid n$. Furthermore, by Fermat's Theorem $d \mid p - 1$. Since p is the smallest prime divisor of n, then $d = 1$. This implies that

$$100 \equiv 121 \pmod{p}$$

and, therefore, $p \mid 21$, which leads to a contradiction.

Problem 6 (Combinatorics)

In the beginning there is a box with n danish cookies, where $n \in \mathbb{Z}^+$. Elsa and Carlos are playing the following game. Elsa goes first and eats a positive integer quantity of cookies that is no more than half of the cookies present in the box. Then goes Carlos and eats a positive integer quantity of that is no more than half of the cookies that are present in the box, etc. At each turn the players can only eat no more than half of the cookies left in the box until one of them cannot make a move and loses the game. Who has a winning strategy?

Solution

Answer: for the numbers n of the form $n = 2^k - 1$ Carlos has a winning strategy; for other numbers n Elsa has a winning strategy.

We will prove the following claim.

Claim

The player who has $2^k - 1$ cookies before their move will loose, i.e. $2^k - 1$ is a "loosing position". All other positions are "winning positions".

Proof

We will lead the proof by induction on k.

For the Basis of Induction we will verify that the statement is true for $k = 1$. Since $2^1 - 1 = 1$, then, obviously, the player who has 1 cookie before their move will loose, which makes it a "loosing position".

For the Inductive Step we will assume that player who has $2^1 - 1 = 1$ cookie before their move, will not be able to make a move, which makes it a "loosing position". We will prove that the player who has $2^{k+1} - 1$ cookies will loose, i.e. $2^{k+1} - 1$ is also a "loosing position".

Lets say that the Player A left the Player B exactly $2^{k+1} - 1$ cookies. Since

$$2 \cdot 2^k = 2^{k+1} > 2^{k+1} - 1$$

then the Player B can only eat no more than $2^k - 1$ cookies. We also have that

$$\left(2^{k+1} - 1\right) - \left(2^k - 1\right) = 2^{k+1} - 2^k = 2^k$$

This implies that the Player B will leave the Player A some number of cookies x, such that

$$2^k \leq x \leq 2^{k+1} - 2$$

Notice that for these values of x we have

$$1 \leq x - 2^k + 1 \leq 2^k - 1$$

Since

$$\frac{2^{k+1} - 2}{2} = \frac{2 \cdot \left(2^k - 1\right)}{2} = 2^k - 1$$

then the Player A will be able to eat any number of cookies from the interval

$$\left[1, 2^k - 1\right]$$

Let us consider the value y defined as $y = x - 2^k + 1$. From above we have that the Player A can eat y cookies on their turn. Now if the Player A indeed eats exactly y cookies, then they will leave the Player B with

$$x - y = x - \left(x - 2^k + 1\right) = 2^k - 1$$

cookies and by the inductive assumption again will put B on the "loosing position". The claim is proven.

We will continue the solution by doing the following casework:

- If $n = 2^k - 1$, then Carlos will win. Indeed, since Elsa goes first, then she will eat at most $2^{k-1} - 1$ cookies and Carlos will be able to leave Elsa with $2^{k-1} - 1$ cookies, which is a "loosing position".

- If $n \neq 2^k - 1$, then Elsa will win. Indeed, since n belongs to some interval of the form $\left[2^k, 2^{k+1} - 1\right]$, then Elsa should eat $n - 2^k + 1$ cookies and will leave Carlos with $2^k - 1$ cookies, which is a "loosing position". ∎

CHAPTER 7

SENIOR LEVEL EXAM OF 2023

Problem 1 (Combinatorics)

Jennifer wrote several pairwise distinct natural numbers on the board and calculated all sums of pairs of different numbers. It turned out that among the last digits of all these sums, all digits from 0 to 9 are present. What could be the smallest number of numbers that Jennifer wrote?

Solution

Answer: the smallest number of numbers is 6.

Start by noticing that for 6 numbers, it is enough to take 1, 2, 3, 4, 5, and 9.

Now we will prove that for five numbers such an example does not exist. Let the five numbers be
$$b_1, b_2, b_3, b_4, b_5$$
Then there will be a total of 10 pairwise sums:
$$b_1 + b_2, b_1 + b_3, \ldots, b_4 + b_5$$

This implies that among them the last digits from 0 to 9 appear exactly once. However, if we add all the pairwise sums we get the number
$$4 (b_1 + b_2 + b_3 + b_4 + b_5)$$
which is even.

Junior, Intermediate and Senior Math Olympiads
by Roman Kvasov, Ph.D.

169

On the other hand, among the pairwise sums, there are exactly 5 of them that end in

$$0, 1, 2, 3, 4, 5, 6, 7, 8, 9$$

so the result of the addition should be odd. We obtained a contradiction.

We conclude that smallest number of numbers that Jennifer could have written is 6, as desired.

Problem 2 (Geometry)

Point M is chosen inside the triangle ABC, such that $\angle BMC = 90^\circ + \frac{1}{2}\angle BAC$. Line AM contains the circumcenter of the triangle BMC. Show that the lines BM and CM contain the circumcenters of the triangles AMC and AMB respectively.

Solution

The solution presented below refers to Figure 7.1.

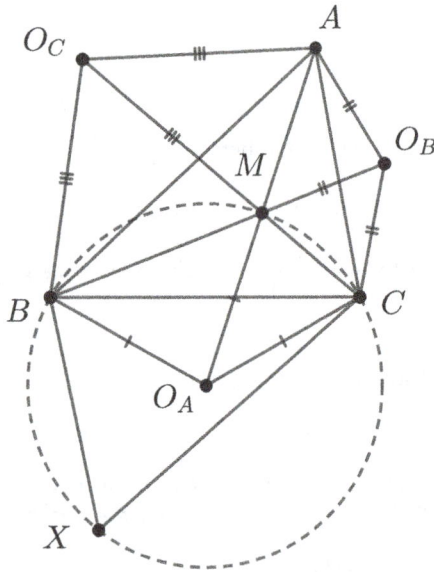

Figure 7.1 M is the incenter of the triangle ABC in Problem 2.

Let O_A, O_B, O_C be the circumcenters of the triangles BMC, AMC, AMB respectively and let the angles of the triangle ABC be equal α, β, γ.

We will show that M is the incenter of the triangle ABC. Let X be any point on the circumcircle of the triangle BMC in the same semiplane as O_A with respect to the line BC. We have

$$\angle BO_AC = 2\angle BXC$$
$$= 2\left(180° - \angle BMC\right)$$
$$= 2\left(90° - \frac{1}{2}\angle BAC\right)$$
$$= 180° - \alpha$$

which implies that the quadrilateral ABO_AC is cyclic.

Since $\angle MO_AC$ is central and $\angle MBC$ is inscribed, we have

$$\angle MO_AC = 2\angle MBC$$

Therefore
$$\angle ABC = \angle AO_AC = 2\angle MBC$$

and MB is the angle bisector of $\angle ABC$. Similarly, MC is the angle bisector of $\angle ACB$ and M is the incenter of the triangle ABC.

From here

$$\angle AMO_B + \angle AMB = \frac{1}{2}\left(180° - \angle AO_BM\right) + 180° - \angle BAM - \angle ABM$$
$$= 90° - \frac{\beta}{2} + 180° - \frac{\alpha}{2} - \frac{\gamma}{2}$$
$$= 180°$$

and the points O_B, M, B lie on the same line.

Similarly, the points O_C, M, C lie on the same line.

Problem 3 (Algebra)

Let a, b, c be positive real numbers. Prove that

$$\frac{a^2 + 2b^2}{2a + b} + \frac{b^2 + 2c^2}{2b + c} + \frac{c^2 + 2a^2}{2c + a} \geq a + b + c$$

Solution

Let us put

$$A = \frac{a^2}{2a+b} + \frac{b^2}{2b+c} + \frac{c^2}{2c+a}$$

$$B = \frac{b^2}{2a+b} + \frac{c^2}{2b+c} + \frac{a^2}{2c+a}$$

Then the left-hand side can be rewritten as follows:

$$\left(\frac{a^2}{2a+b} + \frac{b^2}{2b+c} + \frac{c^2}{2c+a} \right) + 2\left(\frac{b^2}{2a+b} + \frac{c^2}{2b+c} + \frac{a^2}{2c+a} \right) = A + 2B$$

It will be enough to prove that

$$A + 2B \geq a + b + c$$

By applying the Titu's Lemma to the numbers (a, b, c) and $(2a+b, 2b+c, 2c+a)$, and the tuples (b, c, a) and $(2a+b, 2b+c, 2c+a)$, we have

$$A = \frac{a^2}{2a+b} + \frac{b^2}{2b+c} + \frac{c^2}{2c+a} \geq \frac{(a+b+c)^2}{3(a+b+c)} = \frac{a+b+c}{3}$$

$$B = \frac{b^2}{2a+b} + \frac{c^2}{2b+c} + \frac{a^2}{2c+a} \geq \frac{(a+b+c)^2}{3(a+b+c)} = \frac{a+b+c}{3}$$

Consequently

$$A + 2B \geq \frac{a+b+c}{3} + \frac{2(a+b+c)}{3} = a+b+c$$

as desired.

Problem 4 (Geometry)

Line w is drawn through the orthocenter H of an acute triangle ABC. Let w_A, w_B and w_C be the reflections of w across the sides BC, AC and AB, respectively. Prove that w_A, w_B and w_C are concurrent.

Solution

The solution presented below refers to Figure 7.2.

Without loss of generality let us assume that the line w intersects the sides AB and BC at the interior points M and N respectively.

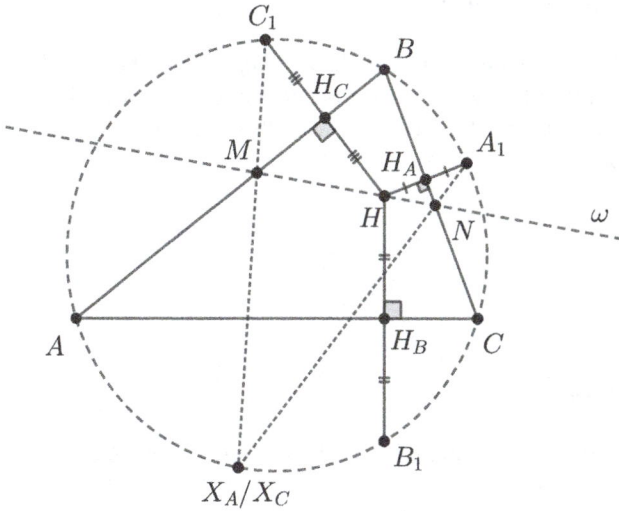

Figure 7.2   Lines ω_A and ω_C intersect on the circumcircle of the triangle ABC in Problem 4.

Let A_1, B_1, C_1 be the reflections of the orthocenter H across the lines BC, AC, AB respectively. Let H_A, H_B, H_C be the bases of the altitudes from A, B, C to the sides BC, AC, AB respectively.

Since A_1, B_1, C_1 belong to the circumcircle of the triangle ABC, then MC_1 is the line ω_C and NA_1 is the line ω_A. Let X_A and X_C be the points of intersection of the lines MC_1 and NA_1 with the circumcircle of the triangle ABC.

We have

$$\angle AA_1X_A = \angle HA_1N = \angle NHA_1$$

and

$$\angle CC_1X_C = \angle HC_1M = \angle MHC_1$$

Since M, H, N are collinear and BH_CHH_A is cyclic, then

$$\angle NHA_1 + \angle MHC_1 = \angle ABC$$

Therefore

$$\angle AA_1X_A + \angle CC_1X_C = \angle ABC$$

and the points X_A and X_C coincide (let us say at the same point X).

This implies that the lines ω_A and ω_C intersect on the circumcircle of the triangle ABC. Since $\angle MHB_1 = \angle XB_1H$, then the line XB_1 is symmetric to ω with respect to AC and ω_A, ω_B and ω_C are concurrent at X.

Problem 5 (Combinatorics)

Numbers 1, 2, ... , n are colored into two colors: blue and red. Determine the smallest n, such that for every coloring of the numbers in two colors, there always exists a *monochromatic* solution (i.e. where all the variables have the same color) of the equation

$$8(x + y) = z$$

Solution

Answer: $n = 2056$.

Let us assume that $n \geq 2056$ and that there are no *monochromatic* solutions of the equation

$$8(x + y) = z$$

Notice that if 1 is blue, then substituting $x = y = 1$ we have that

$$8 \cdot (1 + 1) = 16$$

and the number 16 is red. Furthermore, substituting $x = y = 8$ we have that

$$8 \cdot (16 + 16) = 256$$

and the number 256 is blue. Now substituting $x = 1$ and $y = 256$ we have that

$$8 \cdot (256 + 1) = 2056$$

and the number 2056 is red.

Notice that if the number i is blue then $8 \cdot (i + 1)$ is red, and therefore $i - 15$ cannot be red since then

$$8 \cdot ((i - 15) + 16) = 8(i + 1)$$

This implies that if i is blue, then $i - 15$ is also blue. From here if i is blue, then $i - 15k$ is also blue for all positive integers k, such that $i - 15k > 0$. Since 256 is blue, then we have

$$256 - 15 \cdot 16 = 16$$

and the number 16 should also be blue. We obtained a contradiction.

Now we will show that for $n \leq 2055$ it is possible not to have *monochromatic* solutions. Indeed, let us paint the numbers from the interval $[1, 15]$ in blue, the numbers from the interval $[16, 255]$ in red and the numbers from the interval $[256, 2055]$ in blue. We will proceed by doing the following casework:

- If x, y and z are all colored red, then $16 \leq x, y, z \leq 255$. Therefore

$$z = 8(x + y) \geq 8(16 + 16) = 256 > 255$$

and we obtained a contradiction.

- If x, y and z are all colored blue, then $x, y, z \in [1, 15] \cup [256, 2055]$.

  - If $x, y \in [1, 15]$, then

$$z = 8(x + y) \geq 8(1 + 1) = 16$$

  while

$$z = 8(x + y) \leq 8(15 + 15) = 240$$

  which implies that z is colored red and we obtained a contradiction.

  - If any of the numbers x or y lies in the interval $[256, 2055]$, then

$$z = 8(x + y) \geq 8(256 + 1) = 2056 > 2055$$

  and we obtained a contradiction.

Problem 6 (Number Theory)

Prove that the equation

$$y^2 = x^3 + 7$$

has no integer solutions.

Solution

Let us assume that the given equation has integer solutions.

We will solve this problem by doing the following casework:

- If x is even, then $x = 2k$ for some integer k, and the equation becomes

$$y^2 = x^3 + 7$$
$$y^2 = (2k)^3 + 7$$
$$y^2 = 8k^3 + 7$$

Taking the last equation modulo 4 we have

$$y^2 \equiv 3 \pmod{4}$$

$y \pmod 4$	$y^2 \pmod 4$
0	$(0)^2 \equiv 0$
1	$(1)^2 \equiv 1$
2	$(2)^2 \equiv 0$
3	$(3)^2 \equiv 1$

Let us consider the following table of residues modulo 4:

From here we see that a square of an integer is always congruent to 0 or 1 modulo 4. We obtained a contradiction.

- If x is odd, then let us rewrite the equation as:

$$y^2 = x^3 + 7$$
$$y^2 + 1 = x^3 + 8$$
$$y^2 + 1 = (x + 2)\left(x^2 - 2x + 4\right)$$

Let us prove that the right-hand side of this equation has a prime factor p of the form $4t + 3$. Since x is odd, then the right-hand side of the equation is odd. Let us assume that all prime divisors of $(x + 2)\left(x^2 - 2x + 4\right)$ are of the form $4t + 1$. Then the products of these primes and their powers are also of the form $4t + 1$. This implies that both $x + 2$ and $x^2 - 2x + 4$ are of the form $4t + 1$. However, from

$$x + 2 \equiv 1 \pmod 4$$

we have

$$x \equiv -1 \pmod 4$$

Therefore

$$x^2 - 2x + 4 \equiv (-1)^2 - 2(-1) + 4 \equiv 7 \equiv 3 \pmod 4$$

which leads to a contradiction.

We have shown that the right-hand side of the equation

$$y^2 + 1 = (x + 2)\left(x^2 - 2x + 4\right)$$

has a prime factor p of the form $4t + 3$. Since p divides the right-hand side of the equation, then it should also divide the left-hand side. This implies that -1 is a quadratic residue modulo p. However, -1 is a quadratic residue modulo p only for $p \equiv 1 \pmod 4$. We obtained a contradiction.

CHAPTER 8

SENIOR LEVEL EXAM OF 2024

Problem 1 (Algebra)

Let $x_0 > x_1 > \ldots > x_n$ be real numbers. Prove that

$$x_0 + \frac{1}{x_0 - x_1} + \frac{1}{x_1 - x_2} + \ldots + \frac{1}{x_{n-1} - x_n} \geq x_n + 2n$$

Solution

Let us start by rewriting the given inequality as follows:

$$x_0 - x_n + \frac{1}{x_0 - x_1} + \frac{1}{x_1 - x_2} + \ldots + \frac{1}{x_{n-1} - x_n} \geq 2n$$

Let us introduce the new variables:

$$y_1 = x_0 - x_1$$

$$y_2 = x_1 - x_2$$

$$\ldots \qquad \ldots$$

$$y_n = x_{n-1} - x_n$$

Notice that by the conditions of the problem $y_i > 0$.

Junior, Intermediate and Senior Math Olympiads
by Roman Kvasov, Ph.D.

177

Also, we have

$$y_1 + y_2 + \ldots + y_n = (x_0 - x_1) + (x_1 - x_2) + \ldots + (x_{n-1} - x_n)$$
$$= x_0 - \cancel{x_1} + \cancel{x_1} - \cancel{x_2} + \ldots + \cancel{x_{n-1}} - x_n$$
$$= x_0 - x_n$$

Therefore the given inequality becomes

$$y_1 + y_2 + \ldots + y_n + \frac{1}{y_1} + \frac{1}{y_2} + \ldots + \frac{1}{y_n} \geq 2n$$

Now by AM-GM Inequality for $2n$ variables we have

$$\text{LHS} = y_1 + y_2 + \ldots + y_n + \frac{1}{y_1} + \frac{1}{y_2} + \ldots + \frac{1}{y_n}$$
$$\geq 2n \sqrt[2n]{y_1 \cdot y_2 \cdot \ldots \cdot y_n \cdot \frac{1}{y_1} \cdot \frac{1}{y_2} \cdot \ldots \cdot \frac{1}{y_n}}$$
$$= 2n \sqrt[2n]{\cancel{y_1} \cdot \cancel{y_2} \cdot \ldots \cdot \cancel{y_n} \cdot \frac{1}{\cancel{y_1}} \cdot \frac{1}{\cancel{y_2}} \cdot \ldots \cdot \frac{1}{\cancel{y_n}}}$$
$$= 2n$$

as desired.

Problem 2 (Combinatorics)

A rectangle $m \times n$ with $m \neq n$, consists of mn squares 1×1 and is divided into exactly 8 different polygonal figures following the lines of the grid. Determine the smallest value of mn.

Solution

Answer: 26.

Notice that there exists only 1 figure of area 1, 1 figure of area 2 and 2 different figures of area 3. Therefore, the area of 8 different figures will be at least

$$1 \cdot 1 + 1 \cdot 2 + 2 \cdot 3 + 4 \cdot 4 = 25$$

If the area of the rectangle is equal to 25, then $mn = 25$.

We will proceed by doing the following casework:

- If $m = 5$ and $n = 5$, then we obtain a contradiction with the condition that $m \neq n$.

- If $m = 1$ and $n = 25$, or $n = 1$ and $m = 25$, then all the polygonal figures are simply the rectangles of the form $1 \times a$. Since all figures are different, then the total area should be at least

$$1 + 2 + 3 + 4 + 5 + 6 + 7 + 8 = 36 > 25$$

and we obtained a contradiction.

This implies that the area of the rectangle is at least 26. It is not hard to see that the rectangle 2×13, which has area 26, can be divided into 8 figures of size: 1, 2, 3, 4, 4, 4, 4, 4 (see Figure 8.1).

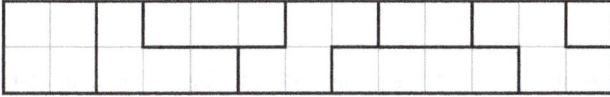

Figure 8.1   Rectangle 2×13 is divided into 8 figures in Problem 2.

Problem 3 (Geometry)

Point K is chosen inside the parallelogram $ABCD$. M is the midpoint of BC and P is the midpoint of KM. Prove that $AK = DK$, if it is know that $\angle APB = \angle CPD = 90°$.

Solution

The solution presented below refers to Figure 8.2.

Let us translate the triangle BPC by the vector $\overline{BA}$ and let Q be the image of the point P. Let L be the midpoint of AD.

Start by noticng that
$$AB \parallel PQ \parallel ML \parallel CD$$
Since $BP \parallel AQ$ and $PC \parallel DQ$, then

$$\angle PAQ = \angle PDQ = 90°$$

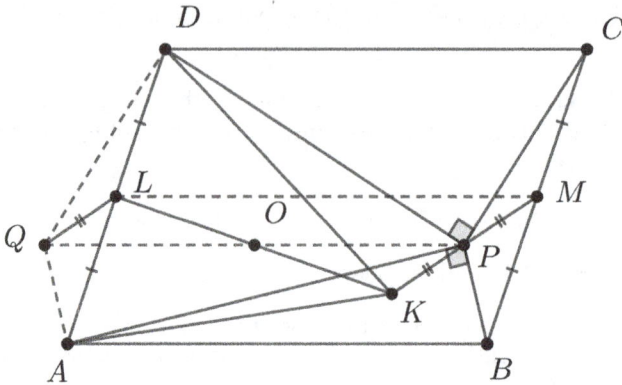

Figure 8.2   Triangle BPC is translated by the vector $\overline{BA}$ in Problem 3.

and the quadrilateral $APDQ$ is cyclic with PQ being the diameter of the circle.

Let O be the intersection of the lines PQ and KL. Let us consider the triangle KML. Since $PO \parallel ML$, and $KP = PM$, then PO is the midsegment of the triangle KML and $2PO = ML$.

Since $ML = PQ$, then O is the midpoint of PQ and therefore is the center of the circle of the quadrilateral $APDQ$.

From here it follows that $OL \perp AD$, and, therefore $KL \perp AD$ and $AK = DK$, as desired.

Problem 4 (Algebra)

For the real numbers a, b, c, the following condition holds:

$$a^2 + c - bc = b^2 + a - ca = c^2 + b - ab.$$

Does it necessarily follow that $a = b = c$?

Solution

Answer: yes.

Let us rewrite the first equality as follows

$$a^2 + c - bc = b^2 + a - ca$$
$$a^2 - b^2 + ac - bc = a - c$$
$$(a - b)(a + b) + c(a - b) = a - c$$
$$(a - b)(a + b + c) = a - c$$

Similarly we obtain

$$(b - c)(a + b + c) = b - a$$

and

$$(c - a)(a + b + c) = c - b$$

We will proceed by doing the following casework:

- If $a + b + c = 0$, then we have

$$a - c = 0$$
$$b - a = 0$$
$$c - b = 0$$

  which implies that in this case all numbers are equal.

- If any two numbers are equal, for example, $a = b$, then from the equality

$$(a - b)(a + b + c) = a - c$$

we have that $a = c$, which implies that in this case all numbers are equal.

- If all numbers are pairwise distinct, and their sum is not zero, then let us multiply all the obtained equalities:

$$(a - b)(b - c)(c - a)(a + b + c)^3 = (a - c)(b - a)(c - b)$$
$$(a - b)(b - c)(c - a)(a + b + c)^3 = -(a - b)(b - c)(c - a)$$
$$\cancel{(a - b)(b - c)(c - a)}(a + b + c)^3 = -\cancel{(a - b)(b - c)(c - a)}$$
$$(a + b + c)^3 = -1$$
$$a + b + c = -1$$

However, this implies that

$$b - a = a - c$$
$$c - b = b - a$$
$$a - c = c - b$$

From here
$$a - c = b - a = c - b$$
and consequently, $a = b = c$, which implies that in this case all numbers are equal.

We conclude that it does necessarily follow that $a = b = c$, as desired.

Problem 5 (Number Theory)

Are there three distinct positive integer numbers a, b, c, such that

$$2a + \text{lcm}(b, c) = 2b + \text{lcm}(a, c) = 2c + \text{lcm}(a, b)$$

Solution

Answer: there are no such numbers.

Let us assume that three such distinct positive integer numbers exist and without loss of generality assume that $a > b > c$.

Start by noticing that $a \mid \text{lcm}(a, b)$ and $a \mid \text{lcm}(a, c)$.

Now let us rewrite the equations as

$$2(b - c) = \text{lcm}(a, b) - \text{lcm}(a, c)$$
$$2(a - b) = \text{lcm}(a, c) - \text{lcm}(b, c)$$
$$2(a - c) = \text{lcm}(a, b) - \text{lcm}(b, c)$$

Since the left-hand side of the first equation is divisible by a, then $a \mid 2(b - c)$. Similarly we have that $b \mid 2(a - c)$ and $c \mid 2(a - b)$.

However, since
$$2a > 2b > 2(b - c) > 0$$
then $2(b - c) = a$. This implies that $b \mid 4b - 6c$ and $c \mid 2b - 4c$. From here $b \mid 6c$ and $c \mid 2b$.

Let $6c = b \cdot k_1$ and $2b = c \cdot k_2$. Multiplying these equations we have

$$k_1 \cdot k_2 = 12$$

This implies that k_1 can only be 1, 2, 3, 4, 6 or 12.

Notice that the values $k_1 = 1, 2, 3$ do not work. For $k_1 \geq 6$, we have

$$6c = b \cdot k_1 \geq 6b$$

This implies that $c \geq b$. We obtained a contradiction.

We conclude that there are no such numbers.

Problem 6 (Combinatorics)

$2k$ stones are divided into 3 not necessarily equal piles. We can take any pile with even number of stones and transfer exactly half of the stones into any other pile. Is it possible to achieve a pile with exactly k stones?

Solution

Answer: yes.

Notice that if we can reach the configuration of the form $(x, 2x, y)$, then from here we can obtain $(x, x, x + y)$ and then $(x, \frac{3x+y}{2}, \frac{x+y}{2})$. Here the middle pile will contain half of all stones.

Now let us show that it is possible to obtain the configuration of the form $(x, 2x, y)$ by the following process. Take a pair of piles, such that at least one of them contains an even number of stones, say $(2m, n)$.

We will proceed by doing the following casework:

- If $m = n$, then we obtained the needed configuration.

- If $m \neq n$, then we will transform this couple, so that one of them always contains an even number of stones and the total number of stones in the couple either always decreases or either always stays the same, but the value $m - n$ decreases. From here we have:

  - If m and n are even then we take m stones and put them into the third pile.

  - If m and n are odd then we take m stones and put them into the second pile to obtain $(m, m + n)$. In this case $m + n$ is even and

    $$\left| \frac{m + n}{2} - m \right| = \left| \frac{n - m}{2} \right| < |m - n|$$

    and eventually m will become equal n.

CHAPTER 9

TOPICS FOR SENIOR MATH OLYMPIADS

Bernoulli's Inequality

Bernoulli's Inequality states that for $x > -1$ and $n \geq 1$ the following inequality holds:

$$(1 + x)^n \geq 1 + nx$$

while for $x > -1$ and $0 < n \leq 1$ the following inequality holds:

$$(1 + x)^n \leq 1 + nx$$

The equality holds when $x = 0$ or $n = 1$.

Jensen's Inequality

Jensen's Inequality states that if the function $f(x)$ is concave upwards on the interval (a, b), then for the real numbers x_1, x_2, ..., x_n from (a, b):

$$\frac{f(x_1) + f(x_2) + \ldots + f(x_n)}{n} \geq f\left(\frac{x_1 + x_2 + \ldots + x_n}{n}\right)$$

If the function $f(x)$ is concave downwards on the interval (a, b), then for the real numbers x_1, x_2, ..., x_n from (a, b):

$$\frac{f(x_1) + f(x_2) + \ldots + f(x_n)}{n} \leq f\left(\frac{x_1 + x_2 + \ldots + x_n}{n}\right)$$

Junior, Intermediate and Senior Math Olympiads
by Roman Kvasov, Ph.D.

185

Rearrangement Inequality

Given real numbers $(a_1, a_2, \ldots, a_n)$ and $(b_1, b_2, \ldots, b_n)$, such that $a_1 \leq a_2 \leq \ldots \leq a_n$ and $b_1 \leq b_2 \leq \ldots \leq b_n$. Let $(c_1, c_2, \ldots, c_n)$ be some permutation of $(b_1, b_2, \ldots, b_n)$. **Rearrangement Inequality** states that

$$S \geq P \geq R$$

where **sorted sum** S, **reversed sum** R and **permuted sum** P are defined as

$$S = a_1 b_1 + a_2 b_2 + \ldots + a_n b_n$$
$$R = a_1 b_n + a_2 b_{n-1} + \ldots + a_n b_1$$
$$P = a_1 c_1 + a_2 c_2 + \ldots + a_n c_n$$

Chebyshev's Inequality

Given real numbers $(a_1, a_2, \ldots, a_n)$ and $(b_1, b_2, \ldots, b_n)$, such that $a_1 \leq a_2 \leq \ldots \leq a_n$ and $b_1 \leq b_2 \leq \ldots \leq b_n$. **Chebyshev's Inequality** states that

$$S \geq A \geq R$$

where **sorted sum** S, **reversed sum** R and **averaged product** A are defined as follows:

$$S = a_1 b_1 + a_2 b_2 + \ldots + a_n b_n$$
$$R = a_1 b_n + a_2 b_{n-1} + \ldots + a_n b_1$$
$$A = \frac{1}{n} (a_1 + a_2 + \ldots + a_n) (b_1 + b_2 + \ldots + b_n)$$

Proof by Infinite Descent

Proof by Infinite Descent is a technique that establishes the impossibility of a certain condition or property by showing that it leads to an infinite chain of strictly decreasing positive integers. It relies on the following list of steps:

- Assume that the condition holds for some positive integer.

- Show that the property also holds for even a smaller positive integer.

- Observe that there cannot be an infinitely many decreasing positive integers.

- Conclude that the original assumption is false.

Fermat's Little Theorem

Let a be an integer and p be a prime. If a is not divisible by p, then

$$a^{p-1} \equiv 1 \pmod{p}$$

Euler's Totient Function

Given a positive integer n. **Euler's Totient Function** of the number n is defined as the number of positive integers less than or equal to n that are coprime to n. Euler's Totient Function is usually written as $\phi(n)$.

If $p_1, p_2, \ldots, p_k$ are all the distinct prime factors of n, then

$$\phi(n) = n \cdot \left(1 - \frac{1}{p_1}\right) \cdot \left(1 - \frac{1}{p_2}\right) \cdot \ldots \cdot \left(1 - \frac{1}{p_k}\right)$$

Euler's Theorem

Let a be an integer and n be a positive integer. If a is relatively prime to n, then

$$a^{\phi(n)} \equiv 1 \pmod{n}$$

where $\phi(n)$ is Euler's totient function.

Modular Multiplicative Inverse

Let a be an integer and let $n > 1$ be a positive integer. An integer number a^{-1} is called a **modular multiplicative inverse** of the number a modulo n if

$$a \cdot a^{-1} \equiv 1 \pmod{n}$$

Wilson's Theorem

If $n > 1$ is a positive integer, then n is prime if and only if

$$(n - 1)! \equiv -1 \pmod{n}$$

Dirichlet's Theorem

If a and b are relatively prime positive integers, then there are infinitely many primes of the form

$$ak + b$$

Multiplicative Order

Given the integer a and positive integer n. The **multiplicative order** of a modulo n is the smallest positive integer k such that

$$a^k \equiv 1 \pmod{n}$$

The multiplicative order of an integer a modulo n is usually written as

$$\mathrm{ord}_n(a)$$

Fundamental Theorem of Orders

Let a, m and n be positive integers. If a and n are relatively prime, then $\mathrm{ord}_n(a)$ exists and the congruence

$$a^m \equiv 1 \pmod{n}$$

holds if and only if $\mathrm{ord}_n(a)$ divides m.

Quadratic Residue and Quadratic Nonresidue

Given a prime number p. Integer number a is called a **quadratic residue modulo p** if there exists an integer x that satisfies the congruence

$$x^2 \equiv a \pmod{p}$$

If such integer x does not exist, then a is called a **quadratic nonresidue modulo p**.

Legendre Symbol

Given a prime number p and an integer number a. The Legendre symbol is defined as:

$$\left(\frac{a}{p}\right) = \begin{cases} 1 & \text{if } a \text{ is a quadratic residue modulo } p \\ -1 & \text{if } a \text{ is a quadratic nonresidue modulo } p \\ 0 & \text{if } a \text{ is divisible by } p \end{cases}$$

Properties of Legendre Symbol

1. **Perfect Square Property**. For prime p and integer a:

$$\left(\frac{a^2}{p}\right) = 1$$

2. **Euler's Criterion**. For prime p and integer a:

$$\left(\frac{a}{p}\right) = a^{\frac{p-1}{2}} \pmod{p}$$

3. **Congruence Property**. For prime p and integers a and b, if

$$a \equiv b \pmod{p}$$

then

$$\left(\frac{a}{p}\right) = \left(\frac{b}{p}\right)$$

4. **Multiplicative Property**. For prime p and integers a and b:

$$\left(\frac{ab}{p}\right) = \left(\frac{a}{p}\right)\left(\frac{b}{p}\right)$$

Law of Quadratic Reciprocity

If p and q are distinct odd prime numbers, then

$$\left(\frac{p}{q}\right)\left(\frac{q}{p}\right) = (-1)^{\frac{p-1}{2}\cdot\frac{q-1}{2}}$$

190   TOPICS FOR SENIOR MATH OLYMPIADS

Nine-Point Circle

Let ABC be a triangle with orthocenter H. Let M_A, M_B, M_C be the midpoints of its sides, H_A, H_B, H_C be the feet of its altitudes and N_A, N_B, N_C be the midpoints of the segments HA, HB, HC, respectively. Then, the points M_A, M_B, M_C, H_A, H_B, H_C, N_A, N_B, N_C lie on the same circle called the **Nine-Point Circle** (see Figure 9.1).

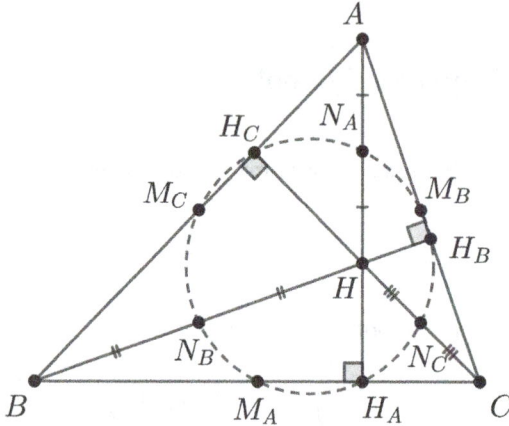

Figure 9.1   Nine-Point Circle of the triangle ABC.

Power of a Point

Given a circle ω with center O and radius R. The **power of the point** X with respect to the circle ω is defined as $d^2 - R^2$, where $d = OX$ (see Figure 9.2).

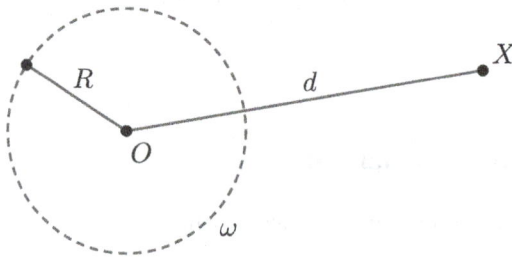

Figure 9.2   Power of the point X with respect to the circle ω is defined as $d^2 - R^2$.

Properties of Power of a Point

Given a circle ω with center O and radius R, and the point X.

1. If the point X lies on the circle ω, then power of the point X with respect to the circle ω is equal to zero.

2. If the point X lies outside the circle ω and a tangent line through X touches ω at A, then the power of the point X with respect to the circle ω is equal to XA^2.

3. If the point X lies outside the circle ω and a secant line through X intersects ω at B and C, then the power of the point X with respect to the circle ω is equal to $XB \cdot XC$.

4. If the point X lies inside the circle ω and a secant line through X intersects ω at B and C, then the power of the point X with respect to the circle ω is equal to $-XB \cdot XC$.

Radical Axis

Radical Axis of two nonconcentric circles is the set of all points whose powers with respect to both circles are equal (see Figure 9.3).

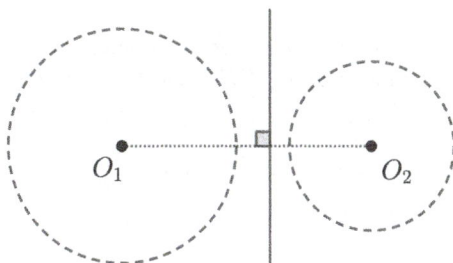

Figure 9.3   The radical axis of two nonconcentric circles is a line perpendicular to O_1O_2.

Radical Center

Let ω_1, ω_2 and ω_3 be the circles with centers at three distinct points O_1, O_2 and O_3, respectively. If O_1, O_2 and O_3 are not collinear, then the three pairwise radical axes of the circles ω_1, ω_2 and ω_3 are concurrent at the point called the **radical center** (see Figure 9.4).

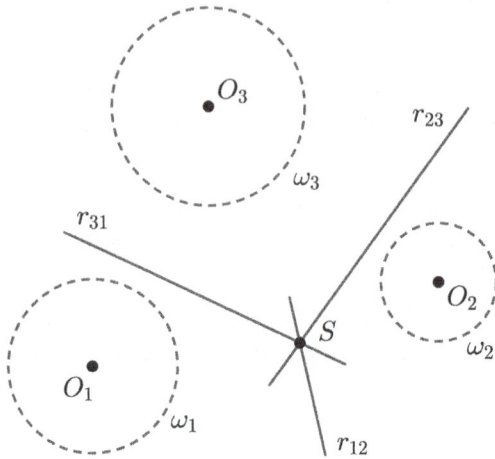

Figure 9.4   The radical axes of the nonconcentric circles ω_1, ω_2 and ω_3 are concurrent at the radical center S.

Ceva's Theorem

In the triangle ABC, let A_1, B_1 and C_1 be the points on the sides BC, AC and AB (see Figure 9.5). The lines AA_1, BB_1, and CC_1 are concurrent if and only if

$$\frac{AC_1}{C_1B} \cdot \frac{BA_1}{A_1C} \cdot \frac{CB_1}{B_1A} = 1$$

Menelaus' Theorem

In the triangle ABC, let A_1 and C_1 be the points on the sides BC and AB, respectively, and B_1 be the point on the extension of the side AC (see Figure 9.6). The points A_1, B_1, and C_1 are collinear if and only if

$$\frac{AC_1}{C_1B} \cdot \frac{BA_1}{A_1C} \cdot \frac{CB_1}{B_1A} = 1$$

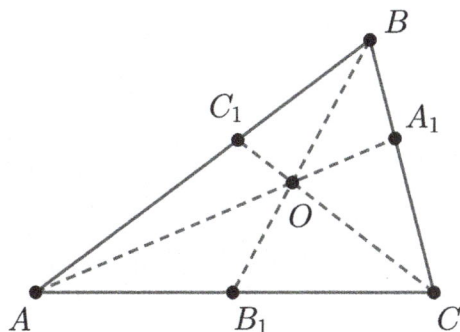

Figure 9.5   Ceva's Theorem for the triangle ABC and the points A_1, B_1 and C_1 on the sides BC, AC and AB, respectively.

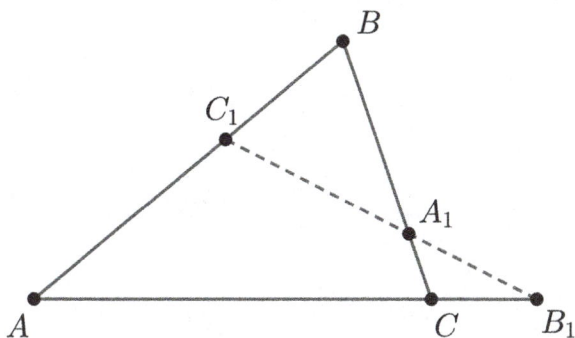

Figure 9.6   Menelaus' Theorem for the triangle ABC, where the points A_1 and C_1 lie on the sides BC and AB, respectively, and the point B_1 lies on the extension of the side AC.

Made in the USA
Monee, IL
08 January 2025

76352774R00108